PROCEEDINGS OF THE INTERNATIONAL CONFERENCE
ON HYDROLOGY AND WATER RESOURCES,
NEW DELHI, INDIA, DECEMBER 1993

VOLUME 2

SUBSURFACE-WATER HYDROLOGY

Water Science and Technology Library

VOLUME 16/2

The titles published in this series are listed at the end of this volume.

PROCEEDINGS OF THE INTERNATIONAL CONFERENCE ON HYDROLOGY AND WATER RESOURCES, NEW DELHI, INDIA, DECEMBER 1993

Volume 2

SUBSURFACE-WATER HYDROLOGY

edited by

VIJAY P. SINGH

Department of Civil and Environmental Engineering, Louisiana State University, Baton Rouge, U.S.A.

and

BHISHM KUMAR

National Institute of Hydrology, Roorkee, India

SPRINGER SCIENCE+BUSINESS MEDIA, B.V.

A C.I.P. Catalogue record for this book is available from the Library of Congress

ISBN 978-94-010-4175-1 ISBN 978-94-011-0391-6 (eBook)
DOI 10.1007/978-94-011-0391-6

Printed on acid-free paper

To

Mother India

TABLE OF CONTENTS

SECTION 3: GROUNDWATER RECHARGE

SECTION 4: AQUIFER RESPONSE 151

SECTION 5: GROUNDWATER RESOURCE EVALUATION

PREFACE

Water is vital to life, maintenance of ecological balance, economic development, and sustenance of civilization. Planning and management of water resources and its optimal use are a matter of urgency for most countries of the world, and even more so for India with a huge population. Growing population and expanding economic activities exert increasing demands on water for varied needs--domestic, industrial, agricultural, power generation, navigation, recreation, etc. In India, agriculture is the highest user of water.

The past three decades have witnessed numerous advances as well as have presented intriguing challenges and exciting opportunities in hydrology and water resources. Compounding them has been the growing environmental consciousness. Nowhere are these challenges more apparent than in India. As we approach the twenty-first century, it is entirely fitting to take stock of what has been accomplished and what remains to be accomplished, and what accomplishments are relevant, with particular reference to Indian conditions.

India has had a long tradition of excellence spanning over several millennia in construction of water resources projects. She has produced civil engineers of highest quality, some of whom rank among the best in the world. In keeping with India's tradition of honoring distinguished engineers at their retirement, it was decided to hold an International Conference on Hydrology and Water Resources, December 20-22, 1993, in New Delhi, to celebrate the retirement of Professor Satish Chandra, Director of the National Institute of Hydrology at Roorkee, who has had a long and distinguished career spanning over three decades in hydrologic education, research, and development. Through this forum, it was hoped to bring together from all over the world researchers, educationists, and practitioners to New Delhi, India, to present the results of their latest work, to share their experiences, to interact with Indian water resources professionals, and to, hopefully, set the stage for future course of hydrology and water resources.

The objectives of this conference were therefore (1) to assess the current state of the art of hydrology and water resources, (2) to discuss applicability of hydrologic and water resources models, (3) to enhance interdisciplinary interaction, (4) to discuss research needs in hydrology and water resources, and (5) to determine the directions for further research.

We received an overwhelming response to our call for papers. The subject matter of the conference was divided into 36 major topics encompassing virtually all facets of hydrology and water resources. Each topic comprised an invited state-of-the-paper and a number of contributed papers. These contributions provided a natural blend to evolve a synthesized body of knowledge on that topic. Extended abstracts of all the invited and contributed papers were assembled in a pre-conference proceedings volume. This helped stimulate discussion and exchange of ideas during the conference.

The papers presented at the conference were reviewed by the Organizing Committee. As a result, some of the papers did not pass the review and were therefore eliminated from inclusion in the final proceedings. The papers contained in this book, SUBSURFACE-WATER HYDROLOGY, represent one part of the conference contributions. The other parts are embodied in three separate companion books, SURFACE-WATER HYDROLOGY, WATER-QUALITY HYDROLOGY, and WATER RESOURCES PLANNING AND MANAGEMENT, which are being published simultaneously. Arrangement of the contributions in these books under four different titles was a natural consequence of the diversity of topics presented in the papers. These books can be treated almost independently, although some overlap may exist between them.

This book contains five sections encompassing major aspects of subsurface-water hydrology. Each section usually starts with invited state-of-the-art papers, followed by contributed papers. Beginning with unsaturated flow in the first section, the papers go on to discuss seepage and drainage in Section 2, groundwater recharge in Section 3, aquifer response in Section 4, and groundwater resource evaluation in the concluding Section 5.

The book will be of interest to researchers as well as those engaged in practice of Civil Engineering, Agricultural Engineering, Hydrology, Water Resources, Earth Resources, Forestry, and Environmental Sciences. The graduate students as well as those wishing to conduct research in water resources planning and management will find this book to be of particular value.

We wish to take this opportunity to express our sincere appreciation to all the members of the Organizing and Advisory Committees, the Louisiana State University administration, and the Indian Association of Hydrologists. Numerous other people contributed to the conference in one way or another, and a lack of space does not allow us to list all of them by name here. We would, nonetheless, like to single out Mr. S. S. Kanwar, who served as Assistant Conference Organizing Secretary, for his inordinate assistance and effort without which the conference would not have come to fruition. Numerous other people contributed to the conference to varying degrees. The authors, including the invited keynote speakers, contributed to the conference technically and made the conference what it was; and this book is largely a result of their efforts. The session chairmen/co-chairmen administered the sessions in a positive and professional manner. We owe our sincere gratitude to all of these individuals.

If the success of a conference is measured in terms of the quality of participants and presentations, then most people would agree that this was a very successful conference. A large number of internationally well-known people, who have long been recognized for their contributions and have long been at the forefront of hydrologic and water resources research, came to participate in the conference. More than 20 countries, covering five continents and most of the countries of the world active in research in hydrology and water resources, were represented. It is hoped that long and productive personal associations and friendships will develop as a result of this conference.

V. P. Singh
Conference Chairman
B. Kumar
Organizing Secretary

ACKNOWLEDGMENTS

The International Conference on Hydrology and Water Resources was sponsored and co-sponsored by a number of organizations. The sponsors supported the conference financially without which it might not have come to fruition. Their financial support is gratefully acknowledged. The co-sponsors extended their support by announcing the conference through their journals, periodicals, transactions, newsletters, or magazines. The publicity so gained helped with increased participation in the conference, and is sincerely appreciated. The following is a list of symposium sponsors and co-sponsors.

Sponsors

Wallac Oy
Finland

Pharmacia United Limited
New Delhi, India

Electrotek Instruments Inc.
Madras, India

Elron Instruments Company Pvt. Ltd.
New Delhi, India

Department of Power
Govt. of Arunachal Pradesh
Itanagar, India

Environmental Measurements & Control
Cochin, India

AIC Agro Instruments (P) Ltd.
Calcutta, India

Spectrum Associates
New Delhi, India

Anubhav Printers & Packers
Roorkee, India

Ajay Printers
Roorkee, India

Co-Sponsors

American Geophysical Union
USA

International Association for
Hydraulic Research
The Netherlands

International Water Resources Association
USA

Soil and Water Conservation Society
USA

Association of American Geographers
USA

International Statistical Institute
The Netherlands

SECTION 1:

UNSATURATED FLOW

VISUALISATION OF STATIC AND DYNAMIC WATER PHENOMENA IN SOIL USING MAGNETIC RESONANCE IMAGING

M.H.G. AMIN*, L.D. HALL+, R.J. CHORLEY*,
K.S. RICHARDS*, T.A. CARPENTER+ & B.W. BACHE*

*. *Department of Geography, University of Cambridge, Downing Place, Cambridge CB2 3EN, UK.*
+. *Herchel Smith Laboratory for Medicinal Chemistry, University of Cambridge School of Clinical Medicine, Robinson Way, Cambridge CB2 2PZ, UK.*

ABSTRACT. This paper reports the use of Nuclear Magnetic Resonance Imaging (MRI) to study static and dynamic soil water phenomena. Two-dimensional and three-dimensional images were obtained to visualise water content, void geometry, and water wetting, drainage and redistribution in three soil materials: packed crumbs of Hanslope clay soil, medium quartz sand and coarse quartz sand. The observed wetting and drying phenomena differed among these soil materials. The wetting front in a column of the medium sand was the typical piston type, but not in the packed clay soil column. Wetting front instability was observed in the coarse sand column. Images of water draining vertically through the medium sand column showed a uniform drainage profile. Drainage from the coarse sand column initially showed a drying front, and approached a uniform profile after free drainage. In contrast, drainage from the clay soil column showed a drying front and a non-uniform drainage profile. The preliminary results from this study show that MRI appears to be a useful tool for visible, nondestructive investigations of soil structure and transport processes, with potential attendant benefits for quantification and modelling of preferential flow in such soils.

1. Introduction

Static and dynamic water behaviour involving storage and flow in soils is important for the understanding and management of hillslope hydrological and hydrochemical processes. A basic technical challenge in such soil-water investigations is the need for visible and non-destructive monitoring of spatial and temporal water distribution, as a prelude of its quantitative analysis. In other areas it has already been demonstrated that Nuclear Magnetic Resonance Imaging (MRI) allows direct visualisation and non-destructive measurements of both static and dynamic water phenomena (Callaghan, 1991). Although some reports have been published on the use of MRI for soil water content measurement and on the effect of soil properties on MR image quality (Anderson & Gantzer, 1989; Bottomley et al., 1986; Brown et al., 1991; Gummerson et al., 1979; Omasa et al., 1985; Rogers & Bottomley, 1987; Tollner et al., 1991), a wider range of applications of MRI for soil-water studies still needs to be developed, particularly in respect of the movement of soil water. Previous work by the authors (Amin et al., 1993a) has shown that experimental methods and parameters can be identified which permit satisfactory MR images to be obtained for soils with modest iron content. The present work presents a demonstration of MRI for the visualisation of soil water content, void geometry, wetting, draining and water redistribution in soil.

MRI of soil-water phenomena depends on the MR signal arising from the protons of the

V. P. Singh and B. Kumar (eds.), Subsurface-Water Hydrology, 3–16.

hydrogen atoms in water. The relative intensity within an MR image is determined by soil MR properties, and by the experimental method and parameters that are used. For a spin-echo pulse sequence (SE), the amplitude of the MRI signal and hence the intensity of a particular point in an image is given by the equation (Kean & Smith, 1986):

$$I = I_0 \,.\, [1 - \exp(-TR/T_1)] \,.\, \exp(-TE/T_2)$$

where:
I is signal amplitude or intensity of point on the image; I_0 is proportional to proton density; TR is repetition time between 90^0 pulses; TE is time to echo; T_1 is spin-lattice relaxation time; and T_2 is spin-spin relaxation time.

This equation states the effect of the characteristic MR properties (I_0, T_1 and T_2) and of experimental parameters (TR and TE) on the final SE image intensity. For a given set of experimental parameters, the regions in the image containing large amounts of proton (i.e. high water content) appear light; the regions with long T_1 appear dark and those with short T_2 appear dark. Thus, by appropriate choice of TR and TE, MR images of wet soil can be made to represent spatial maps of water distribution and therefore delineate the structure of the soil system. Since MRI is a non-destructive technique, time-controlled sequential images can be used to monitor changes in image density which reflect spatial and temporal water transport processes (Amin et al., 1993b).

2. Materials And Methods

The topsoil of a Hanslope series clay from a woodland near Cambridge, England was used for the present study. This contained about 27 g kg^{-1} of iron oxides and about 156 g kg^{-1} of organic matter. Sieved soil crumbs in the size range 1-2 mm (dry bulk density = 0.96 g cm^{-3}) were used. For comparison, sieved medium quartz sand (0.212-0.355 mm equivalent diameter, 1.61 g cm^{-3} dry bulk density) and coarse quartz sand (0.5-1.0 mm equivalent diameter, 1.53 g cm^{-3} dry bulk density) with about 1 g kg^{-1} content of iron oxides were also employed. The samples were packed in perspex cylinders (54 mm inner diameter). The heights of the soil columns varied from 40 mm to 70 mm for different experiments.

For the water drainage experiments, soil or sand columns (54 mm diameter and 60 mm height) were contained on the top of a gauze filter in a perspex cylinder, the lower part of the cylinder under the gauze functioning as a water reservoir. The samples were initially saturated from the bottom of the column until a layer of ponded water appeared on top. Water was allowed to drain slowly through a tube in the bottom of the cylinder under stopcock control. For the wetting experiments, the oven-dried soil samples were contained in a similar perspex cylinder so that air and infiltrating water could escape from the bottom of the container. In addition, a nylon gauze was placed on the top of the samples, and above that a perforated perspex disk was placed to supply the step-feed water evenly and to protect the soil surface from disruption. For other experiments, the samples were packed in perspex cylindrical cups. Distilled water was added from a pipette to the top of the soil columns before or during the experiments according to the individual experimental requirements.

The MR images were obtained using an Oxford Research Systems Biospec I spectrometer operating at 84 MHz for protons, connected to an Oxford Instruments 310-mm horizontal bore, 2 Tesla, superconducting magnet. Laboratory-designed gradient coils produced three orthogonal linear magnetic field gradients (Carpenter et al., 1989). The samples were located in a single-turn solenoid transverse resonator constructed in the laboratory, with an inner diameter of about 60 mm and a height of about 90 mm (Hall et al.,1985). The image-processing, -display and -measurement were conducted via a purpose-built network of UNIX workstations, operating under UNIX-C software written by members of the laboratory.

All images were obtained using a spin-echo (SE) spin-warp (SW) imaging pulse sequence. The sweep width used was 100 kHz and the read gradient was 11 kHz cm^{-1}. For two-dimensional (2-D) imaging, a prefocused 90^0 slice-selective pulse (Roberts et al., 1990) and a 180^0 non-selective pulse were used; the former excited a region of 10-mm thickness through the samples. A 256 x 256 matrix of complex data (consisting of a real and imaginary value) was acquired and no signal average was required. The spatial resolution was about 0.35 mm. The echo time (TE) was 4.46 ms and the repetition time (TR) was 1 s. Total time for acquiring one image was 264 s. For three-dimensional (3-D) imaging, a non-selective 90^0 and a 180^0 pulse were used. A 128 x 128 x 128 data matrix was acquired and no signal averaging was conducted. The spatial resolution was about 0.7 mm. The total acquisition time was about 5 hrs, using TE of 2.91 ms and TR of 1 s.

3. Results And Discussion

Images of vertical central planes (z,y) of packed clay soil columns (54 mm x 40 mm) with different bulk water content are shown in Fig. 1. Figure 1a is an image of the soil with 17% gravimetric water content; b) is the soil with 34%, c) is with 51% water content and d) saturated (~ 70% water content). Although the signal intensity of the image of the soil with 17% water content (Fig. 1a) is so weak that the image is hardly seen, for these samples the numerical value of the signal intensity of the images increases with the increase of water content. Despite the fact that the water was added to the soil columns 30 hrs before the imaging acquisition, the water distribution was not uniform in the soil column, as is especially clear in Fig. 1c. Uniform water distribution can be achieved by using smaller soil columns and allowing a longer redistribution time. This is important for two reasons. First, an even water distribution is necessary for quantitative calibration of the relation between soil-water content and imaging-signal intensity. Second, it is possible to use such a relation from MRI to study the rate of water movement within a sample of carefully controlled properties. The latter is discussed below.

To demonstrate the feasibility of using MRI for visualisation of void geometry, three-dimensional (3-D) images of a column (54 mm x 70 mm) packed clay soil were acquired (Fig. 2). A perspex rod (5 mm diameter) and a piece of perspex sheet (1.6 mm thick and 15 mm wide) were buried in the soil column before the soil saturation, and the image was acquired about 8 hrs after the saturation. Figure 2a shows a 3-D image of the soil column by exposing the thirty-seventh horizontal slice from the top (about 16 mm) and the thirty-seventh vertical slice from the front (about 16 mm). Horizontal cross-sections of the buried rod and sheet can be seen as dark voids in the image, analogous to empty pores or cracks without a fill of water. The dark areas in the image arise because perspex does not contain water and therefore does not give an MRI signal. Figure 2b was obtained by revealing the 3-D image from the thirty-seventh slice from the top and the forty-seventh slice from the front (about 23 mm); both vertical and horizontal cross sections of the rod and sheet appear as dark voids in the image. It can be seen from Fig. 2 that the shapes and sizes of the buried objects are delineated. A 3-D image data set has the advantage that the image can be viewed from arbitrary orientations and slices, which is particularly advantageous when the voids to be visualised are random in size and distribution.

Figures 3, 4 and 5 show the images of zy central planes of wetting and water redistribution of a packed medium quartz sand, a coarse quartz sand and a clay soil column (54 mm x 60 mm), respectively. In each case an oven-dried soil column was first imaged; no signal was visible in those images and hence they are not presented here. Then 20 ml of water was added in one step to the top of the soil column. Figures 3a, 4a and 5a were obtained 5 min after that water addition, and the images of water redistribution were obtained 27 min (Figs. 3b and 4b) or 9.5 min (Fig. 5b) after the first feeding of water. Subsequently a further 10 ml of water was added to the column, and the images (Figs. 3c, 4c, and 5c) were obtained within 5 min. Figures 3d and 4d were obtained 27 min and Fig. 5d were 9.5 min after that second addition of

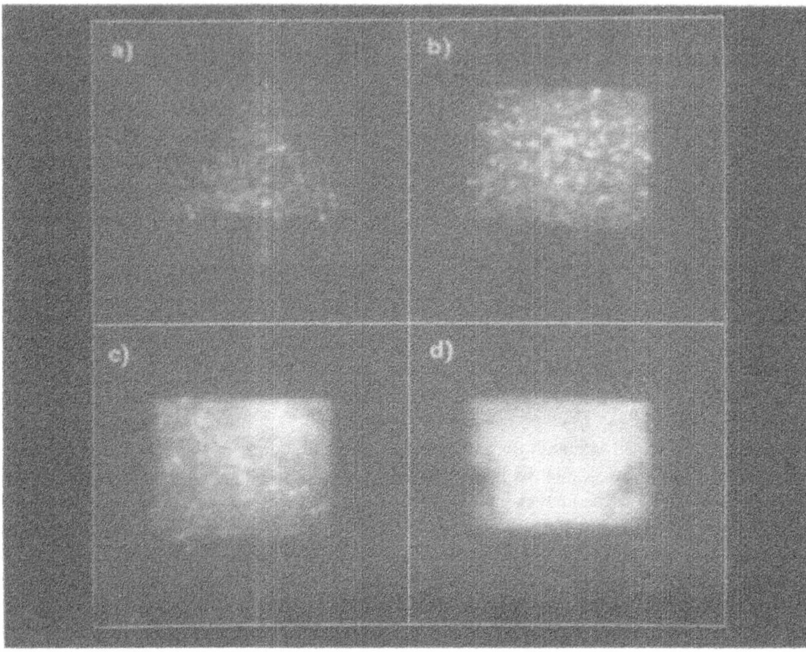

Fig. 1. Two-dimensional (z, y) slice images a)-d) of packed clay soil columns with 54 mm diameter and 40 mm height: a) an image of the soil with 17% gravimetric water content, b) with 34%, c) with 51% water content and d) saturated (~ 70% water content). These illustrate that qualitatively the signal intensity of MR imaging increases with the increase of water contents.

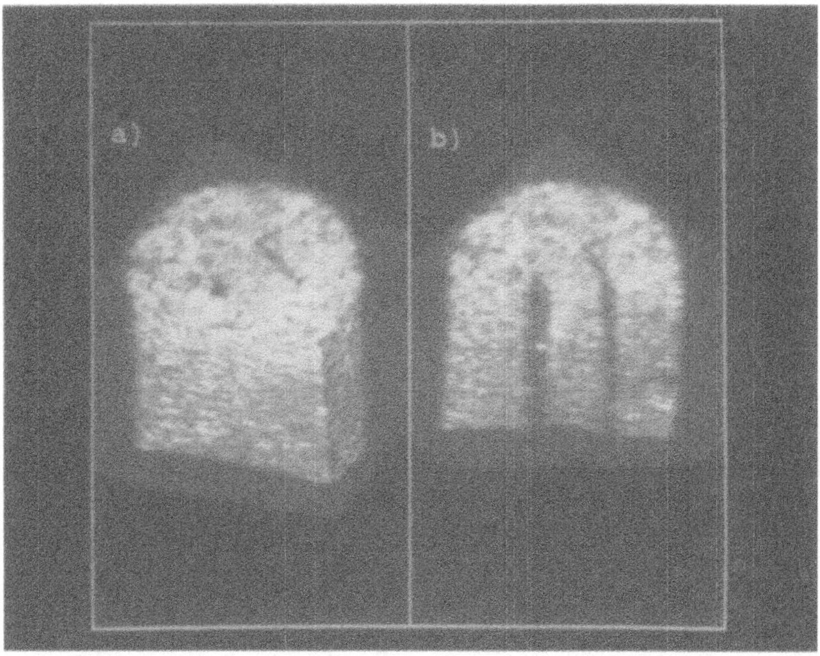

Fig. 2. 128 x 128 x 128 three-dimensional images of a packed clay soil column (54 mm diameter and 70 mm height) containing a perspex rod (5 mm diameter) and a piece of perspex sheet (1.6 mm thick and 15 mm wide): a) is a 3-D overview of the column with a view of the 37th horizontal (z, x) slice from the top, which shows the cross-section of the rod and sheet; in b) the 3-D image has been sectioned to display the 37th slice from the top and the 49th slice from the front, which shows the length of the rod and a section through the sheet.

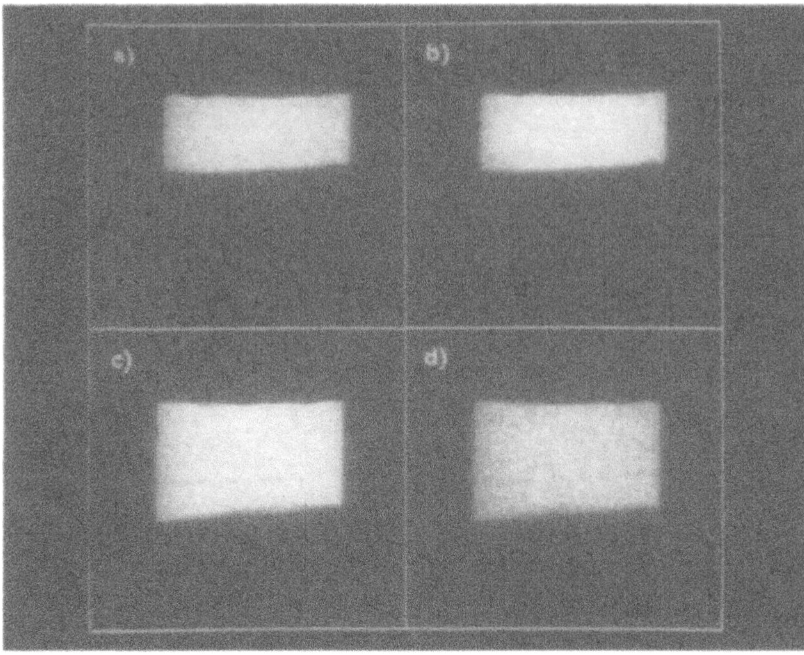

Fig. 3. Two-dimensional (z, y) slice images a)-d), of a packed medium quartz sand column (54 mm diameter and 60 mm height). These show the piston-type wetting and water redistribution processes: a) water distribution in the column 5 min after step-feeding of 20 ml water; b) 27 min after the first water addition; c) 5 min after another addition of 10 ml water; and d) 27 min after the second feeding.

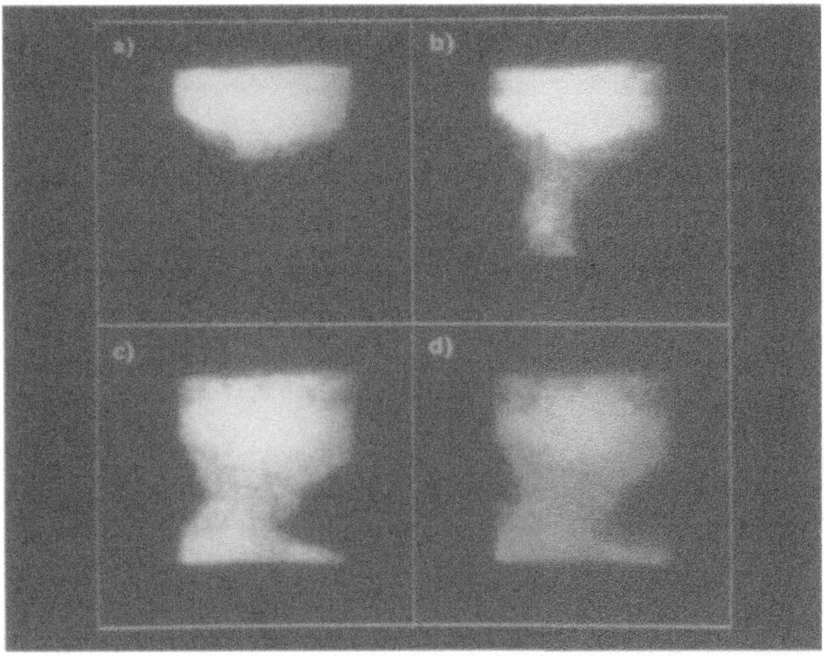

Fig. 4. Two-dimensional (z, y) slice images a)-d) of a packed coarse quartz sand column (54 mm diameter and 60 mm height). All the experimental details are the same as in Fig. 3, and the images demonstrate the wetting and water redistribution through a preferential pathway or a "finger" initiated by instability at the wetting front.

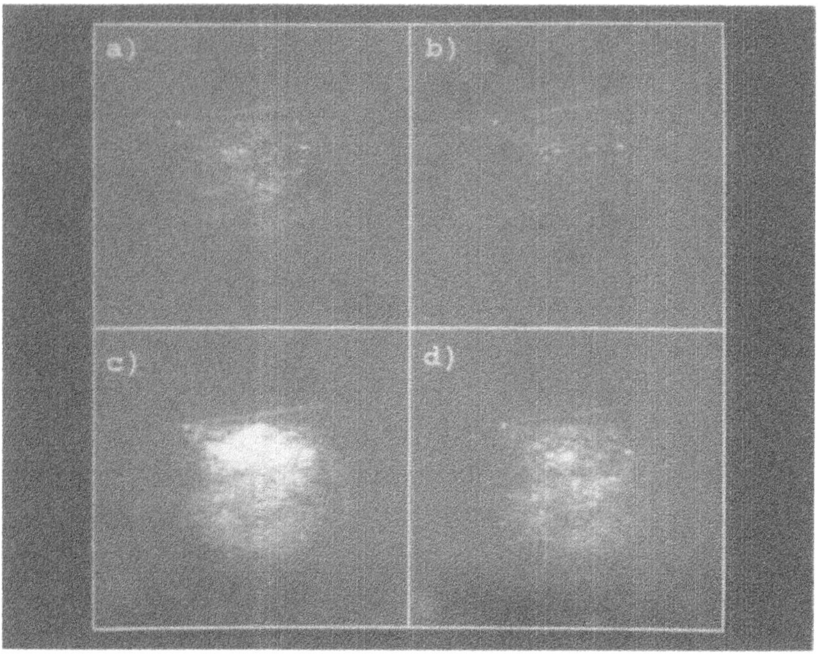

Fig. 5. Two-dimensional (z, y) slice images a)-d) of a packed clay soil column (54 mm diameter and 60 mm height): a) water distribution in the column 5 min after step-feeding of 20 ml water; b) 9.5 min after the first water addition; c) 5 min after another addition of 10 ml water; and d) 9.5 min after the second feed. The images show non-piston type wetting and water redistribution.

water. These images revealed different phenomena of water wetting and redistribution for the three different samples.

The wetting front in the medium sand (Fig. 3) was of the piston type (Bouma, 1981) and was stable during each of the periods of water redistribution. However, it is interesting that a zone of low signal intensity which coincides with the region of the wetting front from the first step-feeding of water can be seen after the second addition of water (Figs. 3c and 3d). This needs further investigation as it seems to be due to trapped air inside the pores. It appears that the loss of signal intensity is due to both a decrease of water content and a decrease of local T_2 values due to a susceptibility effect (Callaghan, 1991).

The wetting front in the coarse sand after the first step-addition of water was curved (Fig. 4a) and there are clear signs of wetting to slightly greater depth at the centre of the sample. Wetting-front instability (Hillel, 1987) was revealed in Fig.4b during the subsequent period, in that redistribution occurred towards the centre of the sample, clearly demonstrating water movement through a preferential pathway. The subsequent wetting of the column from the second feeding of water was unstable and water infiltrated down the column through the expanded "finger" (Fig. 4c). A small amount of drained water was subsequently found inside the tube in the bottom of the container, even though the sample was not saturated (Figs. 4c and 4d). Despite the fact that the mechanism of wetting-front instability is not completely understood (Hillel, 1987), experimental studies by Diment and Watson (1985) suggest that the instability is more likely to occur in coarse-grained sand. The wet finger of soil which results from the unstable wetting-front migration after the first addition of water then probably favours rapid localized infiltration following the second, because of the higher hydraulic conductivity in this zone. Further investigation of this phenomenon is in progress by the group.

The wetting in the packed clay soil column appeared to be neither a piston type nor involved fingering (Fig. 5). The straight dotted line above the soil in Fig. 5 is the signal from water retained by the gauze. It can be seen, by comparing Fig. 5a with Figs. 3a and 4a, that water infiltrated deeper into the soil column than into the sand ones. It seems probable that infiltrated water by-passed the top soil through inter-aggregate pores during the first infiltration period under the experimental conditions. This could be the reason why the signal intensity was much lower in Fig. 5a than in Figs. 3a and 4a, although susceptibility influences also contributed. Another phenomenon can also be noted, by comparing Fig. 5 with Figs. 3 and 4, that the signal attenuation between Fig. 5a and 5b, and between 5c and 5d is greater than that between the corresponding pairs in Figs. 3 and 4, even though the time interval between the pairs in figure 5 is shorter. This is probably due to the water redistribution from soil inter-aggregate pores to intra-aggregate pores and due to the swelling of the clay soil. Both of these phenomena would reduce the T_2 relaxation time of the image regions and therefore significantly attenuate the image signal intensity. The mechanisms of these phenomena need further investigation based on quantitative analysis of the relationships between water content and T_1 and T_2, which are beyond the scope of this general survey.

Figs. 6, 7 and 8 are time-sequential images of water draining vertically from the central plane of a packed column (54 mm x 60 mm) of medium quartz sand, of coarse sand and of clay soil, respectively. In each case the column was initially saturated with a layer of ponded water on the top, and then the water was allowed to drain slowly through a tube in the bottom of the cylinder under the control of a stopcock. Figures 6a, 7a and 8a were obtained before the beginning of drainage, and the other images were taken at 4.5 min intervals. Figures 6d, 7d and 8d were acquired after the water had drained freely with the stopcock fully opened. These images reveal different patterns of water drainage from the three soil columns under the experimental conditions. Drainage from the clay soil column had a drying front, although this was not regularly shaped (Fig. 8). Drainage from the coarse sand column (Fig. 7) also had a drying front when the sand was drained from saturation initially (Fig. 7b), and approached a uniform drained profile finally (Fig. 7d). In contrast, the water distribution during drainage from

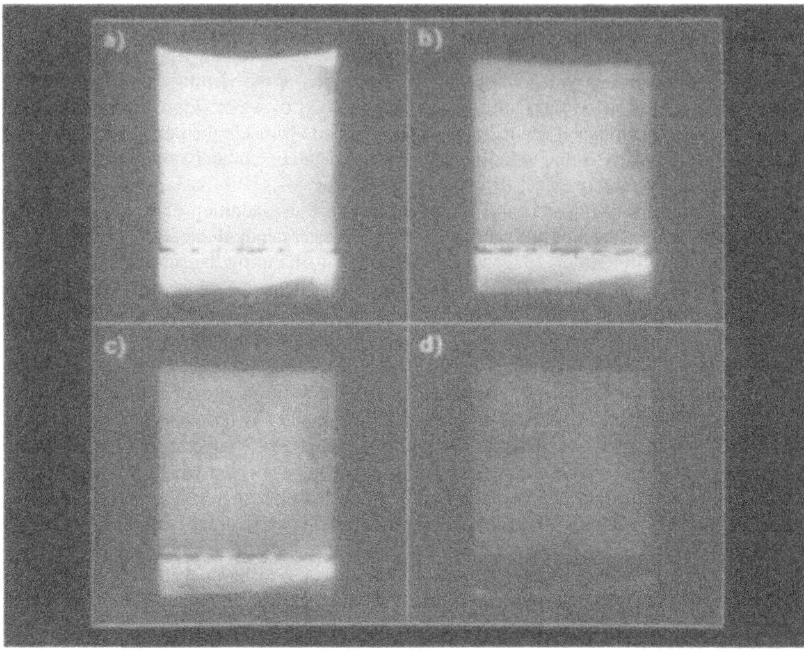

Fig. 6. Two-dimensional (z, y) slice images a)-d), of a packed medium quartz sand column (54 mm diameter and 60 mm height). These illustrate water draining vertically from the column: a) an image of the central plane of the column before the beginning of drainage; b) 5 min after and c) 9.5 min after the beginning; and d) after drainage following full opening of the stopcock.

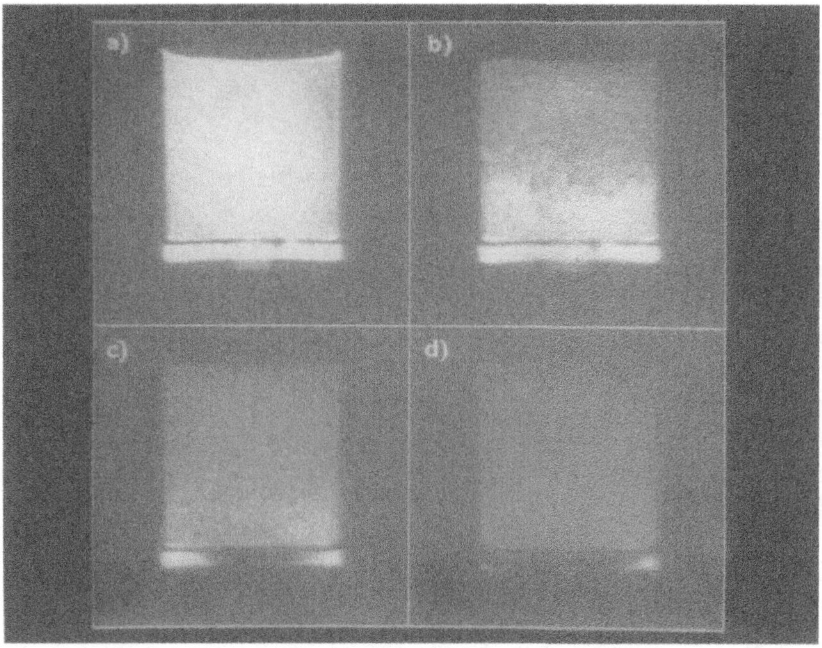

Fig. 7. Two-dimensional (z, y) slice images a)-d) of a packed coarse quartz sand column. All experimental details are the same as in Fig. 6.

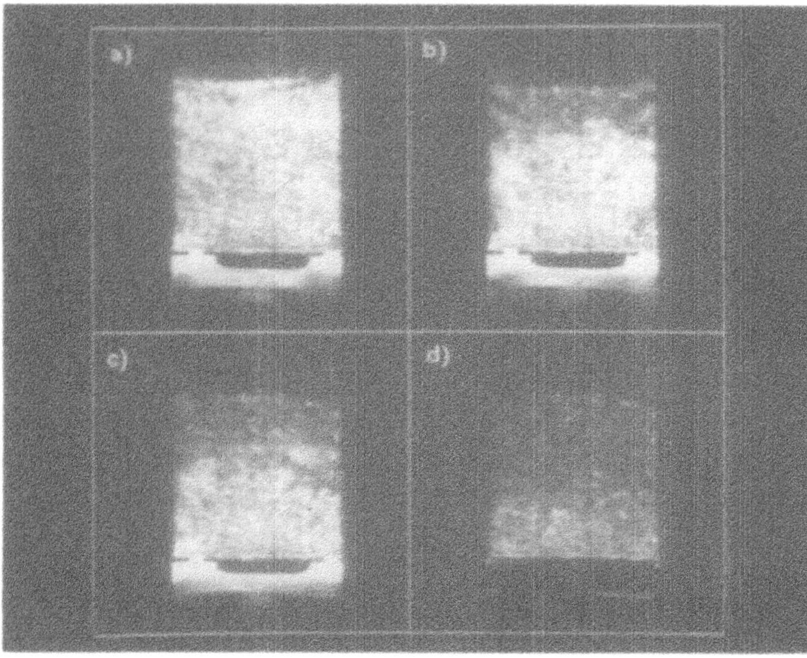

Fig. 8. Two-dimensional (z, y) slice images a)-d), of a packed clay soil column (54 mm diameter and 60 mm height). These illustrate water draining vertically from the column: a) an image of the central plane of the column before the beginning of drainage; b) 4.5 min after and c) 9 min after the beginning of drainage; and d) after drainage following full opening of the stopcock.

the medium sand column was spatially and temporally uniform (Fig. 6). It seems that the water distribution during draining reflects different influences in the three materials. In the medium sand column drainage was dominated by surface tension effects, and therefore the moisture profile was uniform. In the coarse sand column it was influenced by both gravity and surface tension when near saturation, and was dominated by the latter after free drainage. Finally, in a packed clay soil column drainage was non-uniform, probably due to a wide range of pore sizes and a wide range of water contents held within different pores in the soil column.

4. Conclusions

The results presented in this paper are reproducible under the experimental conditions. Although the results are only qualitative at this stage, they demonstrate the feasibility of MRI methods for a wide range of soil water studies. Quantitative measurements of water content, infiltration, drainage and water redistribution in soil, especially soil with macropores, can be carried out using MRI by careful calibration and experimental control and will be reported in subsequent papers.

In the interim, a number of conclusions can be drawn from this study.

1. MRI can be used for the study of static and dynamic water phenomena in both "model" soil systems and in natural soils with modest iron content and sufficient water content to provide MRI signal.
2. The uniformly-packed clayey soil studied exhibits a non-piston wetting front due to the wide range of the pore size distribution. In contrast, a packed medium sand column has a piston-type wetting front. "Fingering" of an unstable migrating wetting front is apparent in the coarse sand.
3. During drainage the water distribution in the medium sand column is uniform; that in the soil column is non-uniform but has a drying front; and in the coarse sand column it has a drying front initially, but approaches a uniform profile after it has freely drained.
4. Consequently, MRI appears to be a potentially useful tool for visible, non-destructive measurements of water in soil, having considerable potential for the study of the influences of wetting and drying processes, and of preferential pathways and preferential flow on hydrological and hydrochemical processes.

Acknowledgements

This work was made possible by a munificent benefaction to LDH, TAC and partial financial support to MHGA from Dr Herchel Smith. We thank Dr. S. J. Gibbs at HSLMC and Dr M. Cislerova at Czech Technical University, Prague, Czechoslovakia for helpful discussions and Mr. C. Harbird for his excellent assistance on manufacture of some of the hardware.

References

AMIN, M. H. G., HALL, L. D., CHORLEY, R. J., CARPENTER, T. A., RICHARDS, K. S. & BACHE, B. W. 1993a. 'Spatial and temporal mapping of water in soil by magnetic resonance imaging', *Hydrological Processes*, **7**, (In press).

AMIN, M. H. G., HALL, L. D., CHORLEY, R. J., CARPENTER, T. A., RICHARDS, K. S. & BACHE, B. W. 1993b. 'Magnetic resonance imaging of soil-water phenomena', *Magnetic Resonance Imaging* (In press).

ANDERSON, S. H. & GANTZER, C. J. 1989. 'Determination of soil water content by X-ray

computed tomography and magnetic resonance imaging', *Irrigation Science*, **10**, 63-71.
BOTTOMLEY, P.A., ROGERS, H.H. & FOSTER, T.H. 1986. 'NMR imaging shows water distribution and transport in plant root systems in situ', *Proceedings of the National Academy of Sciences, United States of America*, **83**, 87-89.
BOUMA,. 1991. 'Soil morphology and preferential flow along macropores', *Agricultural Water Management*, **3**, 235-250.
BROWN, D. P., PRATUM, T. K., BLEDSOE, C., FORD, E. D., COTHEN, J. S. & PERRY, D., 1991. 'Noninvasive studies of conifer roots: nuclear magnetic resonance (NMR) imaging of Douglas-fir seedlings', *Canadian Journal of Forest Research*, **21**, 1559-1566.
CALLAGHAN, P. T. 1991. *Principles of Nuclear Magnetic Resonance Microscopy*, Oxford University Press, New York.
CARPENTER, T.A., HALL, L.D. & JEZZARD, P.J. 1989. 'Proton magnetic resonance imaging of solid polymers using instrumentation designed for the liquid state', *Journal of Magnetic Resonance*, **84**, 383-387.
DIMENT, G. A. & WATSON, K. K. 1985. 'Stability analysis of water movement in unsaturated porous materials: 3. Experimental studies', *Water Resources Research*, **21**, 979-984.
GUMMERSON, R.J., HALL, C. & HOFF, W.D. 1979. 'Unsaturated water flow within porous materials observed by NMR imaging', *Nature*, **281**, 56-57.
HALL, L.D., MARCUS, T., NEALE, C., POWELL, B., SALLOS, J. & TALGALA, S.L. 1985. 'Design of volume resonator probes for NMR imaging', *Journal of Magnetic Resonance*, **62**, 525-528.
HILLEL, D. 1987. 'Unstable flow in layered soils: a review', *Hydrological Processes*, **1**, 143-147.
KEAN, D. & SMITH, M. 1986. *Magnetic Resonance Imaging - Principle and Applications*, William Heinemann Medical Books, London.
OMASA, K., ONOE, M. & YAMADA, H. 1985. 'NMR imaging for measuring root system and soil water content', *Environmental Control in Biology*, **23**, 99-102.
ROBERTS, T.P., CARPENTER, T.A. & HALL, L.D. 1990. 'Design and application of prefocused pulses by simulated annealing', *Journal of Magnetic Resonance*, **89**, 595-604.
ROGERS,H.H. & BOTTOMLEY, P.A. 1987. 'In situ nuclear magnetic resonance imaging of roots: influence of soil type, ferromagnetic particle content, and soil water', *Agronomy Journal*, **79**,957-965.
TOLLNER, E. W., VERMA, B. P., MALKO, J. A., SHUMAN, L. M. & CHESHIRE, J. M. Jr., 1991. 'Effect of soil total iron on magnetic resonance image quality', *Communications in Soil Science and Plant Analysis*, **22**, 1941-1948.

HYDROLOGIC MODELLING ACKNOWLEDGING SPATIAL VARIATIONS OF HYDRAULIC CONDUCTIVITY

R. P. Rudra, W. T. Dickinson and R. K. Gupta
Professor Professor Grad. Student
School of Engineering, University of Guelph, Guelph, Ontario , N1G 2W1 (Canada)

ABSTRACT: Physically-based models are used extensively to simulate the infiltration process at the field scale. In general, the models predict outputs considering the deterministic nature of soil hydraulic conductivity and many other input parameters. A prime difficulty with these predictions is likely to be the lack of accountability for spatial variations of many input parameters involved. This paper presents an approach to include the spatial variability of hydraulic conductivity in field-scale infiltration modelling. A model describing the spatial structure of hydraulic conductivity has been developed to generate spatial hydraulic conductivity data which have been incorporated into a Green-Ampt and Mein-Larson infiltration model. The model has a been evaluated to a plot-scale situation under simulated rainfall conditions. The field experiments were conducted with the Guelph Rainfall Simulator at Agriculture Canada facilities in Ottawa, Ontario (Canada). The analysis of the results reveals that the GAML model, combined with a spatial structure for hydraulic conductivity, is capable of simulating the rainfall infiltration process for a spatially variable field soil.

Introduction

Concerns regarding nonpoint source pollution have increased the importance of understanding critical processes related to the quantity and quality of water movement. They have also placed a greater emphasis on the efficient and wise application of hydrologic and nonpoint source pollution modelling tools. The infiltration process is considered an integral part of all hydrologic and nonpoint source pollution models, and equations describing this process are invariably used in the model applications. The Green-Ampt infiltration equation (1911) has been adopted widely in these models. This infiltration equation is based on the physics of the soil water flow process, and utilizes hydraulic parameters related to the flowing medium. Almost all hydraulic parameters have been found to exhibit a range of variability on the field scale (Nielsen et al., 1973; Biggar et al., 1976 and Jensen et al., 1986); and documented studies in the literature have revealed a significant effects of this variability on infiltration and overland flow (Smith et al., 1979 and Bresler et al., 1983). Many research investigators have shown that the accuracy of model outputs can be improved by inclusion of the variable nature of input parameters (Biggar et al., 1976; Rudra et al., 1986 and Jensen et al., 1986). The successful application of such models, therefore, depends on the carefully determination of input parameters, and on the technique used to represent the variable nature of their occurrence.

A possible framework of analysis of variability-related problems is offered by stochastic modelling techniques (Sharma et al., 1983 and Webster, 1985). In such an analysis of an hydraulic property, that property is interpreted as a stochastic variable and is subjected to uncertainty. In the present study, a stochastic technique has been employed to represent the spatial structure of hydraulic conductivity data of a field soil. The validity of the technique is

V. P. Singh and B. Kumar (eds.), Subsurface-Water Hydrology, 17–32.

evaluated by combining it with an infiltration model. Specifically, the study was conducted with the objectives: (1) to investigate and quantify the spatial structure of hydraulic conductivity on a field-scale; and (2) to analyze the infiltration process by using a Green-Ampt and Mein-Larson (1973) model, evaluating the model performance by comparing modelled infiltration with that obtained with the field experiments conducted under simulated rainfall conditions. The data for the study were collected from a series of field-experiments conducted under rainfall simulation conditions. Details of the field experiments, the selected infiltration model, and the technique of hydraulic conductivity analysis are given in the following section.

Study Area and Experimental Details

The study was conducted in a conventionally-tilled field located at the Greenbelt Farm of Agriculture Canada, Ottawa (Canada). The field has had a drainage system since 1974, and has been monitored for ground water investigations through a battery of four observation wells located at nine different sites in the field. A few physical properties determined for the soil in the experimental area are given in Table (I).

Infiltration experiments were conducted in summer 1990 using the Guelph Rainfall Simulator (GRS). The GRS involves a continuous-spray and wide-angle, full jet nozzle to simulate the rainfall events (Tossell et al., 1987). Its hydraulic characteristics are given in Table (II). Experiments were conducted at ten locations within the field (Figure 1) on a battery of five test plots (1 m X 1 m) installed at each location. Physical conditions of the soil surface at the time of experiments are given in Table (III). Each test-plot was prepared by inserting prefabricated frames into the ground to a depth of 5.0 to 7.5 cm. The rainfall was applied to each test-plot at an intensity of 16.02 cm/h, and the runoff was collected at a one minute time interval in a V-shaped trough. The volume of infiltrating water was computed by subtracting the runoff from the rainfall volume of each time interval. Similarly, the instantaneous infiltration rates were calculated as the difference between the rainfall and runoff rates. The data, thus obtained, were analyzed to determine the hydraulic conductivity and instantaneous infiltration rate at each site.

A single infiltration curve for the study area was obtained by averaging the instantaneous infiltration rate curve of each test site. The hydraulic conductivity data were analyzed for their spatial structure, and the GAML infiltration model was employed to simulate the infiltration process under constant rainfall conditions. A detailed procedure of analysis on these aspects of the study are given in the following sections.

Simulation of Infiltration Process by GAML Model

Water movement in a porous media can be described by the Richards infiltration equation. Physically-based approximations of this equation have been shown to provide a good characterization of water infiltration into soils (Mein and Larson, 1973; Smith and Parlange, 1978). The Green and Ampt infiltration equation (1911) is a member of this class of physically-based approximations. This equation uses Darcy's law to estimate the infiltration rate into a deep homogeneous soil having a sharply defined wetting front under a ponded surface condition, and may be expressed as:

TABLE I: Few Physical and Chemical Properties of Soil at Experimental Site.

Depth (cm)	Sand	Silt	Clay	pH	O.M. (%)
0-10	63.85	26.80	9.40	4.9	3.4
10-20	64.46	27.00	8.60	5.2	3.1
20-30	66.31	29.00	4.70	5.4	--
30-40	60.55	30.60	8.80	5.4	--
40-50	44.30	36.30	19.40	6.0	--

TABLE II: Hydraulic Characteristics of Rainfall Simulation System.

Characteristic	Value
Nozzle Diameter	12.7 mm
Nozzle Height Above Plot	1500 mm
Simulated Rainfall Intensity	160.2 mm/h
Test-Plot Size	1m X 1m
Number of Drops	18667
Kinetic Energy Flux Density	0.7570 $J/m^2/s$

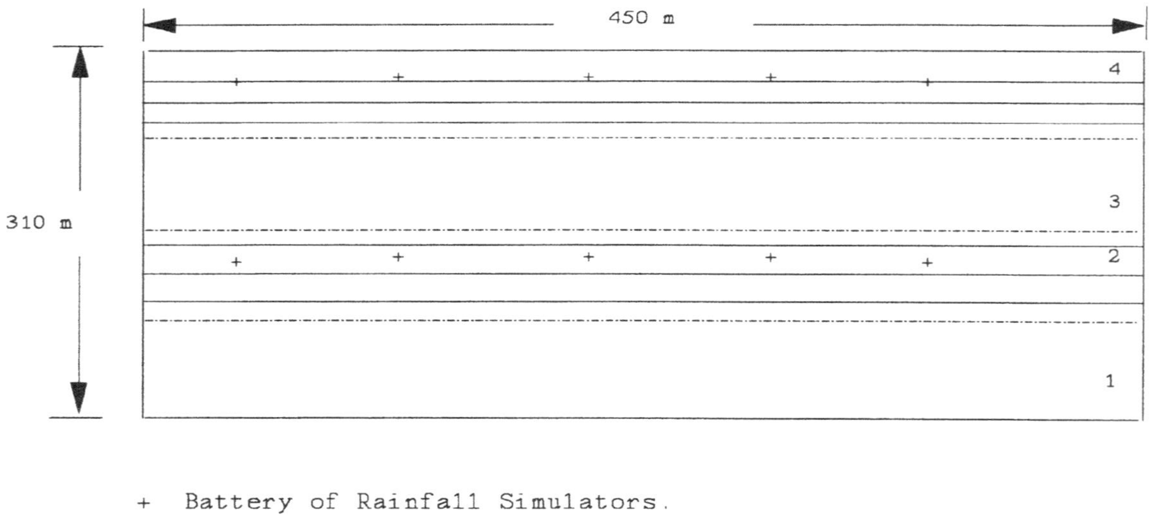

Figure 1. Layout of Rainfall Simulation Experiments at Greenbelt Farm of Agriculture Canada, Ottawa (Canada).

TABLE III: Physical Properties of Soil at Time of Simulation Experiments.

Property	Value
Bulk Density (kg/m^3)	1390-1460
Antecedent Soils Water Content (Vol., %)	20.40
Soil Surface Conditions	Dry and undisturbed
Time in one Experimental Run (min)	60-80

TABLE IV: Statistics of Modelled and Experimentally-Derived Values of Hydraulic Conductivity.

Statistical Parameters	Experimentally-derived values	Model values
Mean Value	0.2156	0.2228
Auto-Corre. Coeff.		
(a) Lag 1	0.2146	0.2562
(b) Lag 2	0.1541	0.1652
(c) Residuals	0.0034	0.0045
Coefficient of Variation	0.2505	0.3213

$$f = K_s\ (\ 1\ +\ S_{av}\ *\ M\ /\ F) \tag{1}$$

where:

- f = infiltration rate, (L/T)
- K_s = hydraulic conductivity of wetted zone, (L/T)
- S_{av} = suction at the wetting front, (L)
- M = initial moisture deficit, (L^3/L^3) and
- F = cumulative infiltration, (L).

Equation (1) can also be expressed in terms of cumulative infiltration rate by substituting $f = dF/dt$, and integrating over the time period t, with the conditions that $F = 0$ at $t = 0$ and $F = F$ at $t > 0$. The solution, thus obtained, may be written as:

$$K_s\ t = F - S_{av}\ M\ *\ \ln\ [1\ +\ F\ /\ (M*S_{av})] \tag{2}$$

where:

- F = cumulative infiltration, (L) and
- t = infiltration time (T).

The basic approach was further extended by Mein and Larson (1973) to account for the general case of delaying surface ponding with a steady rainfall intensity. Their approach included infiltration cases both prior to and after the occurrence of ponding under rainfall conditions. They derived a relationship to predict the volume of infiltration water and the time of ponding for the case of delaying ponding. The relationship is of the form:

$$F_p = M\ S_{av}\ (\ R/K_s\ -\ 1\ ) \tag{3}$$

where:

- F_p = cumulative infiltration at the time of ponding, (L), and
- R = rainfall intensity, (L/T).

Mein and Larson then modified the Green-Ampt equation to predict infiltration after surface ponding by translating the time scale. The modified equation is:

$$K_s(t\ -\ t_p\ +\ t_s)\ =\ F\ -\ MS_{av}\ \ln[1\ +\ F/(M*S_{av})] \tag{4}$$

where:

- t = time from beginning of the rainfall event, (T)
- t_p = time from beginning of the event to ponding, (T), and
- t_s = time required to infiltrate F_p if surface ponding had been instantaneous, (T).

Under rainfall conditions, the infiltration process can be categorized into two stages: the initial stage during which the application rate, R, is lower than the infiltration capacity of the soil, f; and a later stage during which the application rate is greater than the infiltration capacity of the soil. Under these two stages, infiltration may be described by applying Equations (3) and (4) in sequence. The resulting model has been termed the Green-Ampt and Mein-Larson model (Mein and Larson, 1973), and is widely known in the literature as the GAML model. The model has been extended by Tan (1987) to a specific case in which the flow profile is considered to be composed of three soil layers. In the present study, this model has been used for a single layer case.

To apply the GAML model, it is necessary to determine various parameters associated with the rainfall event and the soil profile. The rainfall intensity, R, can be available from either rainfall records or measurements made during infiltration experiments. The soil moisture deficit, M, is the difference in soil water content, on a volumetric basis, before and after the infiltration and may be obtained from soil sampling. The parameters of suction at the wetting front (S_{av}) and saturated hydraulic conductivity (K_s) can be determined by laboratory and field measurements.

The suction at the wetting front is equal to the area under the relative hydraulic conductivity versus matric potential curve with limits of integration from saturation matric potential to initial matric potential. It may be expressed as:

$$S_{av} = - \int_{\psi}^{\psi_i} K_r \, d\psi \tag{5}$$

where:

Ψ_i = capillary suction at initial moisture content, (L),
K_r = relative hydraulic conductivity ($K(\Psi)/K_s$), and
Ψ = saturated value of soil moisture tension, (L).

An exponential relationship between the unsaturated hydraulic conductivity and matric potential was assumed, and S_{av} was obtained by solving the integral of Equation (5). The saturated hydraulic conductivity was determined by conducting rainfall simulation experiments as described in the above section. The data obtained for hydraulic conductivity were analyzed with regard to its variability in the spatial domain.

Spatial Structures of Hydraulic Conductivity

Data series such as experimentally-derived hydraulic conductivity values may be analyzed for their deterministic and stochastic components. These components may be represented through a variety of mathematical models. The deterministic component can often be described using a Fourier Series technique. This technique has been used by Webster (1977) for the periodicity of various soil physical properties; and by Persaud et al., (1983) to model the deterministic and stochastic components of soil temperature in the time domain. In the present analysis, this technique has been employed to describe the deterministic component of hydraulic conductivity in the spatial domain. The series in finite form may be expressed as:

$$X(l) = A_1 + \sum_{l=1}^{N} [\, C_1 \cos(2\pi l/N) + D_1 \sin(2\pi l/N) + C_2 \cos(4\pi l/N) + D_2 \sin(4\pi l/N)\,] \qquad (6)$$

where:

$X(l)$ = the deterministic component of hydraulic conductivity at point l,
N = the spatial sample size of hydraulic conductivity measurements,
A_1 = the mean of the hydraulic conductivity data series,
C_1 and C_2 = model coefficients for cosine terms of the series, and
D_1 and D_2 = model coefficients for sine terms of the series.

The contribution of the first and second harmonics of the deterministic component of the hydraulic conductivity data (i.e. Equation (6)) is revealed by the variance explained by these harmonics. It can be computed as $[(C_1^2 + D_1^2)/2]$ and $[(C_2^2 + D_2^2)/2]$ for the first and second harmonic, respectively. The variance not accounted for by these two harmonics can be attributed to the stochastic component.

Studies have shown that properties measured at positions close to each other may be highly correlated than properties measured at positions further apart (Sharma et al., 1983 and Hopmans, 1989). The correlation can be represented through a variety of auto-regressive models. In the present study, it has been expressed by a second-order auto-regressive scheme of the following form:

$$S(l) = \alpha_1 (S_l - S_{l-1}) + \alpha_2 (S_l - S_{l-2}) + a_l \qquad (7)$$

where:

α_1 and α_2 = the stochastic model parameters for the first and second orders,
S_l = the stochastic component of hydraulic conductivity, and
a_l = the residuals or independent sub-component.

The combination of Equations (6) and (7) yields a model containing deterministic and stochastic components associated with the hydraulic conductivity data. The computational procedure given by Box et al. (1976) was adopted to obtain the auto-regressive model parameters of Equation (7). When all the parameters of Equations (6) and (7) had been determined, the formulated model was employed to generate hydraulic conductivity values for the experimental soil.

The adequacy of the model was examined in two ways: (1) by comparing the statistical characteristics (i.e. mean, coefficient of variation and auto-correlation coefficient) of the generated and experimentally-derived values of hydraulic conductivity, and (2) by checking the residual series, obtained by differencing the generated and experimentally-derived values, for their characteristics of independence.

Results and Discussion

VARIABILITY OF EXPERIMENTALLY-DERIVED HYDRAULIC CONDUCTIVITY

The hydraulic conductivity values obtained from the rainfall infiltration experiments were found to vary from 0.042 to 0.261 cm/min with a mean of 0.158 cm/min and a coefficient of variation of 41.62%. These results indicate that hydraulic conductivity varies widely from one point to an other point across the field. Analysis of the data showed that the auto-correlation coefficients for the hydraulic conductivity data were 0.2146 and 0.1541 at lag 1 and lag 2, respectively. The non-zero values of auto-correlation coefficients suggest that the hydraulic conductivity values are not completely random but are dependent upon their values at adjoining positions.

The above statistical characteristics reveal that hydraulic conductivity values can be expected to exhibit wide variation in the spatial domain, and have mutual dependence on their adjoining values. It suggests that the spatial structure of hydraulic conductivity values can be represented by a combination of deterministic and auto-regressive models.

SPATIAL STRUCTURE OF EXPERIMENTALLY-DERIVED HYDRAULIC CONDUCTIVITY

(1) Deterministic Component: The deterministic component associated with hydraulic conductivity has been expressed by the first and second harmonics of the Fourier Series. It may be expressed as:

$$\mathbf{X(l) = 0.1636 + [\ 0.0253\ Cos\ (\ 2\pi l/N\ ) - 0.0194\ Sin\ (\ 2\pi l/N\ ) - 0.0176\ Cos\ (\ 4\pi l/N\ ) - - 0.0310\ Sin\ (\ 4\pi l/N\ )\ ]} \qquad \mathbf{(8)}$$

The first term of Equation (8) represents the mean of hydraulic conductivity values, the second and third terms represent the first harmonic, and the fourth and fifth terms represent the second harmonic for the hydraulic conductivity. The analysis indicated that the deterministic component accounts for 15% of the total variance. The remaining variance not accounted for by these harmonics is attributed to stochastic variations of the data.

(2) Stochastic Component: Analysis of the stochastic component consisted of a dependent and an independent stochastic components. The dependent component was represented based on the auto-correlation analysis of the data. The correlogram, a relationship between auto-correlation coefficient (ACR) and lag, is shown in Figure (2). This figure shows that the values of ACRs at lag 1 and 2 are 0.2146 and 0.1541, respectively. These values decrease asymptotically with increase in lag except at a few lags. These results indicate that the correlation between two measured values of hydraulic conductivity decreases as the distance between their measuring positions increases. They also reveal that the dependence between two nearby values is considerably greater than between values lying further apart. The stochastic component having such characteristics is expressed by a second-order auto-regressive scheme. The coefficients of the scheme, α_1 and α_2 of Equation (7), were derived from the auto-correlation coefficients at lag 1 and 2, respectively. The stochastic model, thus obtained, may be expressed as:

$$\mathbf{S(l) = 0.4054\ (\ S_l - S_{l-1}\ ) - 0.2616\ (\ S_l - S_{l-2}\ ) + a_l} \qquad \mathbf{(9)}$$

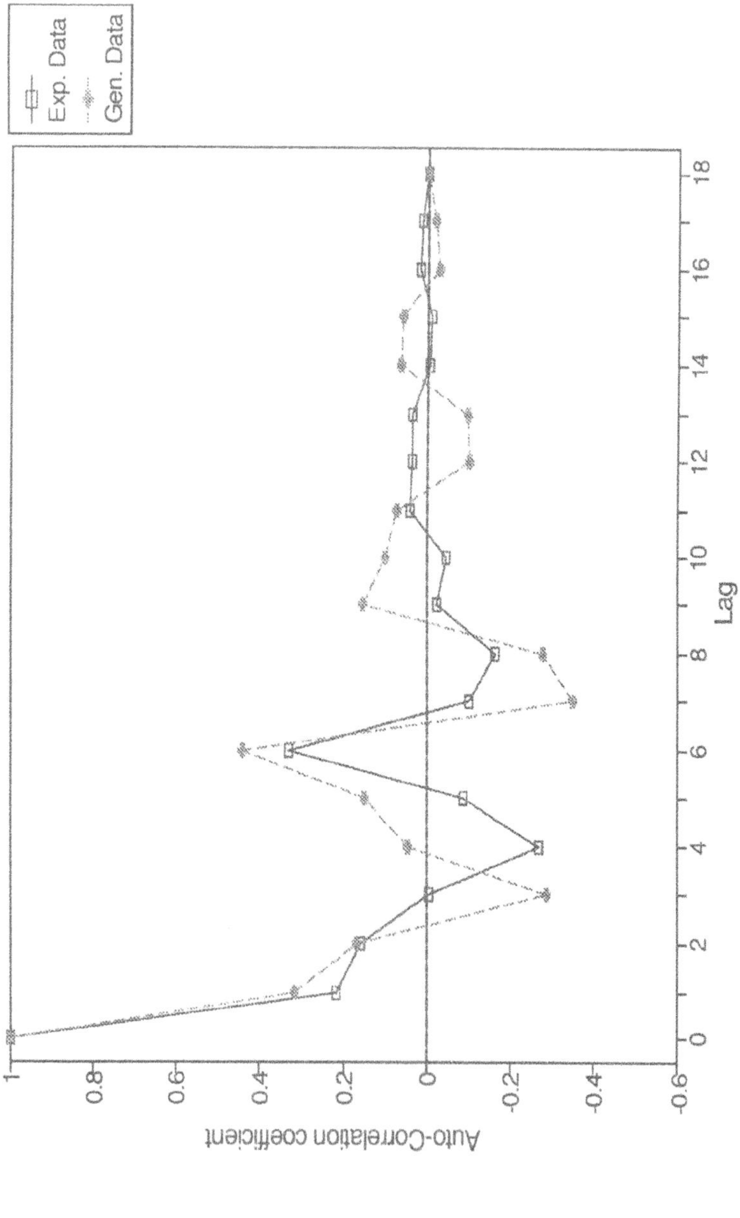

Figure 2. Correlograms of Model and Experimentally-derived Values of Hydraulic Conductivity.

The first and second terms of Equation (9) constitute the dependent part, and the third term represents the independent part of the stochastic component. The analysis indicated that the dependent part accounts for 62% of the total variance explained by the stochastic component of hydraulic conductivity data. The remaining 38% of the variance can not be explained by the auto-regressive model, and is attributed to the independent part of the model. These results are due to the high spatial variability and lower degree of dependence of the hydraulic conductivity on their adjoining values.

The formulated model given by Equations (8) and (9) was employed to generate hydraulic conductivity values at all the measurement points. The generated values were subjected to statistical analysis to test the adequacy of the models. Firstly, the residual series was tested for independence by analyzing its correlogram. The correlogram of residual series, given in Figure (3), shows that the auto-correlation coefficients at the first few lags are small compared to the ACRs of experimentally-derived values of hydraulic conductivity. The coefficients decrease further with lag, and are approximately zero at all lags. These results indicate that the residuals are non-significant in magnitude, and their series may be considered to be independent in nature. Secondly, the mean, coefficient of variation, and ACRs for the generated data were compared with the those of the experimentally-derived values, and are given in Table (IV). They show that the generated values exhibit the same statistical properties as those of the experimentally-derived data. It may, therefore, be considered that the models are capable for representing the spatial variability of the hydraulic conductivity data.

INFILTRATION SIMULATION CONSIDERING SPATIALLY VARIABLE HYDRAULIC CONDUCTIVITY

To simulate infiltration response under rainfall conditions, the generated values of hydraulic conductivity were incorporated into the GAML model. A total of 25 computer runs were made with GAML, and output was obtained at various time steps. The model outputs, in the form of infiltration rates and cumulative infiltration depths, were analyzed with regard to their means and the means $\pm$ standard deviation, and were compared to the infiltration data obtained from the simulated rainfall experiments.

The results given in Figure (4) reveal that the model overpredicts infiltration rate at the first three time steps, and underpredicts for the rest of the time steps in an infiltration run. The higher infiltration values at the initial time steps may be attributed to the fact that the model does not consider crop characteristics (i.e. crop stage, leaf area, and root density and distribution) in simulating the infiltration process. The interception of rain water by crop canopy affects the infiltration rate, particularly at initial time steps, and might have been responsible for the lower infiltration rates during the initial stage of infiltration. However, Figure (4) shows that the observed values fall well within the range of one standard deviation about the mean model values. These results show that the differences between predicted and experimental values are non-significant at the 0.05 level of significance. A comparison based on the cumulative infiltration values reveals a similar observation (Figure (5)). It may, therefore, be inferred that the GAML infiltration model incorporating a spatial structure of hydraulic conductivity is useful for simulating the rainfall infiltration process for a spatially variable soil.

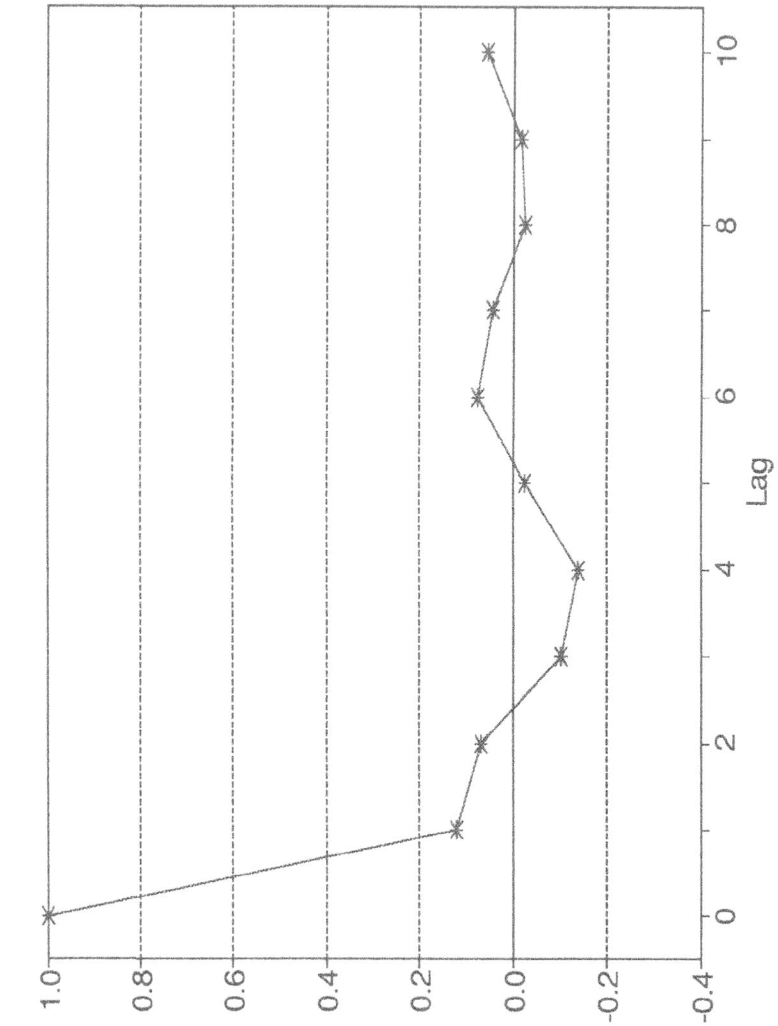

Figure 3. Correlogram of Residual Series of Hydraulic Conductivity Values of Experimental Soil.

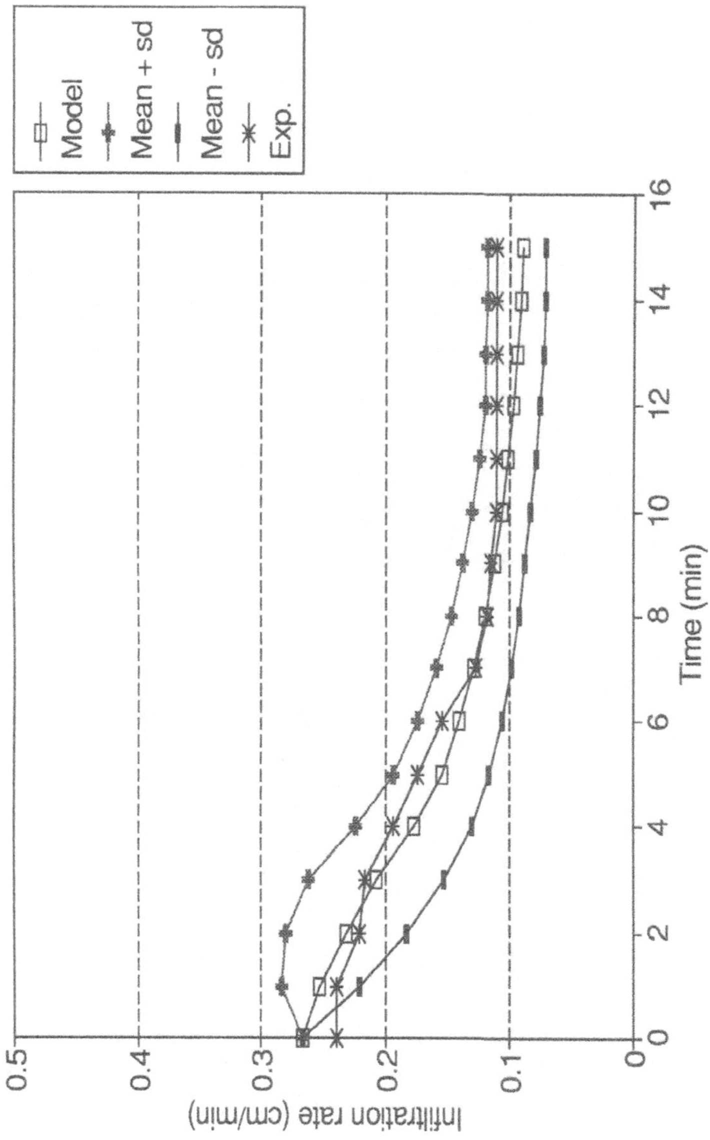

Figure 4. Comparison of Modelled and Observed Values Infiltration Rates of Experimental Soils.

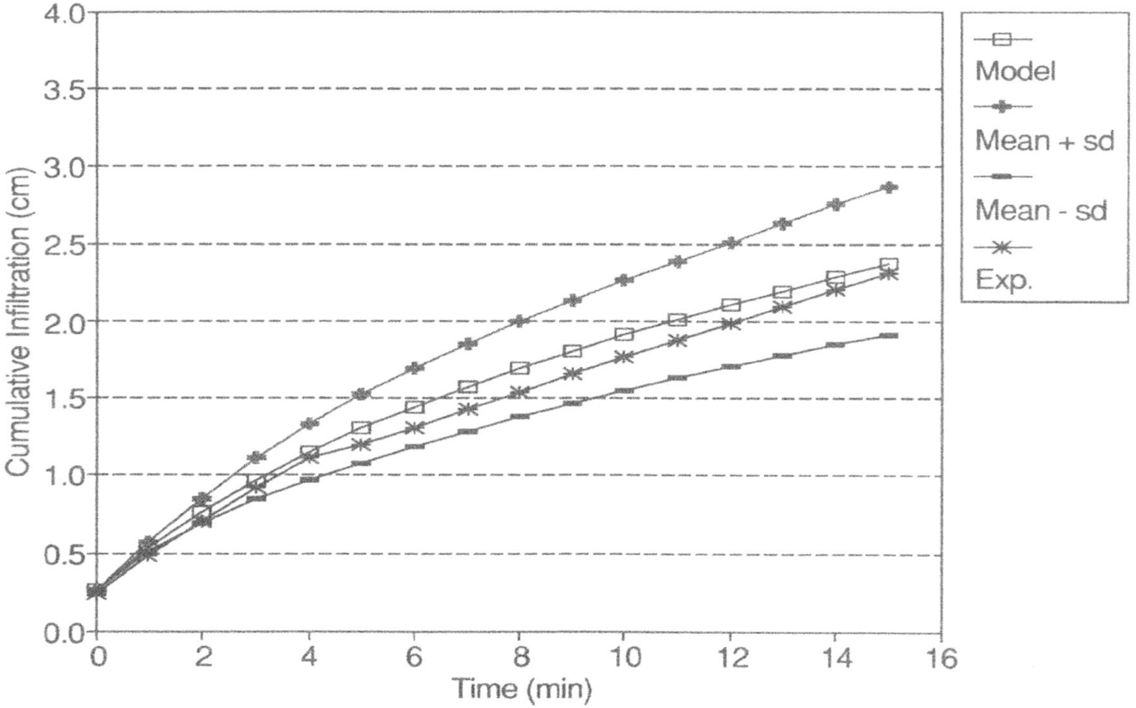

Figure 5. Comparison of Model and Observed Values of Cumulative Infiltration of Experimental Soil.

Summary and Conclusions

A study was conducted to examine the rainfall infiltration process considering a spatially-variable structure for the hydraulic conductivity of a soil. To obtain the data, infiltration experiments were conducted using a rainfall simulator at the Greenbelt Farm of Agriculture Canada, Ottawa (Canada).

The spatial structure of hydraulic conductivity was described by a deterministic-stochastic model. The output of the model was incorporated in the Green-Ampt-Mein-Larson, GAML, model (Tan, 1987) to examine the infiltration response under simulated rainfall conditions. The performance of the infiltration model was evaluated by comparing the modelled and experimental data regarding infiltration rate and cumulative infiltration.

It is concluded from the study that: (1) The hydraulic conductivity exhibits a wide spatial variation across the field. Its value at a point of measurement is correlated with values at adjoining points. The correlation decreases with increase in the spacing between the measurement points. (2) The spatial structure of hydraulic conductivity may be adequately described by a deterministic-stochastic model. (3) The GAML infiltration model, incorporating the spatial structure model of hydraulic conductivity, overpredicts mean infiltration rate at initial stages of infiltration, and underpredicts at later stages. However, the mean values of observed infiltration rates lye within the range of the mean model values $\pm$ one standard deviation. This study indicates that the GAML infiltration model, combined with the spatial model of hydraulic conductivity, is quite useful for simulating rainfall infiltration for a spatially-variable soil.

References

Biggar, R. E. and D. R. Nielsen, 1976. 'The spatial variability of the leaching characteristics of a field soil', Water Resou. Res., **12 : 78 - 84.**

Box, J.E.P. and Jenkins, G.M. 1976. Time series analysis-forecasting and Control, San Francisco, Holden Days, Inc.

Bresler, E. and Dagan, G. 1983. 'Unsaturated flow in spatially variable fields, 2, Application of water flow models to various fields', Water Resou. Res., **19: 421-428.**

Green, W.H. and Ampt, G. 1911. 'Studies of soil physics, Part I- The flow of air and water through soils', J. Agril. Sciences, **4: 1-24.**

Hopman, J.W. 1989. 'Stochastic description of field-measured infiltration data', Trans. of the A.S.A.E., **32(6): 1987-1993.**

Jensen, K.H. and Butts, M.B. 1986. 'Modelling of unsaturated flow in heterogeneous soil', Nordic Hydrology, **281-294.**

Mein, R.G. and Larson, C.L. 1973. 'Modelling infiltration during steady rain', Water Resou. Res., **9(2): 384-394.**

Persaud, N. and Chang, A.C. 1983. 'Estimating soil temperature by linear filtering of measured air temperature', Soil Soc. Soc. Am. J., **47: 841-846.**

Philip, J.R., 1957. 'The theory of infiltration. I. The infiltration equation and its solution', Soil Sci., **83 : 345 - 357.**

Rudra, R. P., Dickinson, W. T., Clark, D. J. and Wall, G. J. 1986. 'Games - A screening model of soil erosion and fluvial sedimentation on agricultural watersheds', Can. Water Resou. J., **11: 58 - 71.**

Sharma, M. L., Barron, R. J. W. and Boer, E. S. De. 1983. 'Spatial structure and variability of infiltration parameters', In: Advances in Infiltration, Proc. National Conference on Advances in Infiltration, December 12-13, **113-121.**

Smith, R.E. and Parlange, J.Y. 1978. 'A parameter efficient hydrologic infiltration model', Water Resou. Res., **14 : 533-538.**

Smith, R. E. and Hebbert, R. H. B. 1979. 'A Monte Carlo analysis of the hydrologic effects of spatial variability of infiltration', Water Resou. Res., **15: 419-429.**

Tan, kim-Ann. 1987. 'Modelling of rainfall infiltration under thawing soil conditions', M. Sc. Thesis, University of Guelph, Guelph, Ontario (Canada), **pp. 164.**

Tossell, R., Dickinson, W.T., Rudra, R.P. and Wall, G.J. 1987. 'A portable rainfall simulator', Can. Agril. Eng., **29: 155-162.**

Webster, R. 1977. 'Spectral analysis of gilgai soil', Aust. J. Soil Res., **15: 191-204.**

Youngs, E. G. 1968. 'An estimation of sorptivity for infiltration studies from moisture movement considerations', Soil Sc., **106 :157 - 163.**

INFILTRATION SUBJECT TO TIME-DEPENDENT SURFACE PONDING: EXACT RESULTS AND CORRESPONDENCE TO SOLUTE TRANSPORT WITH NONLINEAR REACTION

D. A. Barry
Department of Environmental Engineering
Centre for Water Research
University of Western Australia
Nedlands 6009 Western Australia

J.-Y. Parlange
Department of Agricultural and Biological Engineering
Cornell University
Ithaca NY 14853-5701 USA

ABSTRACT. Richards' equation, the governing equation for single phase unsaturated flow in soil, is highly nonlinear and, consequently, is difficult to solve either analytically or numerically. The nonlinearity enters through the soil hydraulic functions, viz., the soil moisture characteristic curve and the unsaturated hydraulic conductivity. We present an exact quasi-analytical solution for the one-dimensional Richards' equation subject to an arbitrary, time-dependent ponding depth. The solution depends on solving a first-order ordinary differential equation. For certain special cases, such as when the surface ponding depth is constant, fully analytical results can be derived. The theory is used in a number of applications. For example, an existing analytical approximation for cumulative infiltration subject to a ponded, time-dependent surface head is checked. As well, a drainage formula is derived. New results for solute transport coupled with a nonlinear adsorption isotherm are derived by making use of an exact mapping between Richards' equation and the governing solute transport equation.

1. Notation

arg	argument
A	function of time
B	constant defined by (54)
c	normalised solute concentration
c_B	adsorption isotherm fitting parameter
c_0	normalised solute concentration at soil surface
c_1, c_2	constants, L
C	function defined by (7), T^{-1}
C_c	value of C when ψ_s is constant, T^{-1}
D	solute diffusion/dispersion coefficient, L^2T^{-1}
D_r	cumulative drainage, L
$Ei(.)$	exponential integral with argument .
g	magnitude of the gravitational acceleration, LT^{-2}
$H(.)$	Heaviside function with argument .
I	cumulative infiltration, L
I_c	quantity of solute in the soil profile, L
K	hydraulic conductivity, LT^{-1}
K_s	saturated hydraulic conductivity, LT^{-1}
n	time step number
p	fluid pressure offset such that $p = 0$ is 1 atmosphere, $ML^{-1}T^{-2}$
s	normalised solid phase concentration
S	sorptivity, $LT^{1/2}$
S_0	sorptivity for zero surface head, $LT^{1/2}$

V. P. Singh and B. Kumar (eds.), Subsurface-Water Hydrology, 33–48.

t	time, T
v	pore fluid velocity, LT^{-1}
$W(.)$	function, with argument ., defined by (19)
z	position, L
z_f	solute front position, L
z_I	initial height of reservoir, L
β	constant, L
γ	constant, L
δ	parameter defined by (27)
ε	constant, T^{-1}
Δt	time increment in numerical scheme to compute A, T
$\Delta\theta$	$\theta_s - \theta_r$
$\Delta\psi$	$\psi_s(0) - \psi_a$, L
θ	volumetric water content
θ_r	residual volumetric water content
θ_s	saturated volumetric water content
μ	parameter defined by (26)
ρ	density of soil water, ML^{-3}
ψ	pressure head, L
ψ_a	air-entry head, L
ψ_s	surface pressure head, L

2. Introduction

The prediction of infiltration into soil is a key element of hydrologic catchment modelling. Green and Ampt (1911) presented a remarkably simple infiltration formula for the cumulative amount of water to enter a dry soil profile due to a ponded surface condition. Their formula is still used widely because it captures the essential short and long time behaviour of ponded water infiltration. Thus, it is of sufficient accuracy for many applications (e.g., Onstad et al., 1973).

The Green-Ampt formula relies on the assumption of a sharp front of infiltrating water. Recently, Barry et al. (1993b) showed that the Green-Ampt formula can be derived without this assumption. Instead, the soil moisture characteristic curve is a particular function of the hydraulic conductivity. It is this latter, less prescriptive, relationship which leads to the Green-Ampt infiltration formula.

The Green-Ampt approach is easily extended to other, more complicated scenarios. For example, a depth-dependence of the hydraulic conductivity or the initial water content in the soil profile can be introduced. A common extension is to consider the ponding depth at the soil surface to be time dependent as it would be, for instance, if runoff or evaporation occurred. In the following, the exact approach of Barry et al. (1993b) is extended to derive an exact solution for the case of a time-dependent surface condition. In addition, a corresponding exact solution is presented for solute transport subject to a nonlinear sorption isotherm.

3. Exact Solution for Infiltration with a Time-Dependent Surface Head

The governing equation for one-dimensional, unsaturated fluid flow is (Richards, 1931)

$$\frac{\partial\theta}{\partial t} = \frac{\partial}{\partial z}\left[K(\psi)\frac{\partial\psi}{\partial z} - K(\psi)\right], \tag{1}$$

where K is the hydraulic conductivity and ψ (= $p/\rho g$) is the pressure head. The z-axis is assumed to be pointed downwards. Recalling that if only infiltration or drainage is considered, then the soil moisture characteristic curve, $\psi(\theta)$, is single-valued, and (1) can be written as

$$\frac{d\theta}{d\psi}\frac{\partial\psi}{\partial t} = \frac{dK}{d\psi}\frac{\partial\psi}{\partial z}\left(\frac{\partial\psi}{\partial z} - 1\right) + K\frac{d^2\psi}{dz^2}. \tag{2}$$

Following Barry et al. (1993b), a solution for (2) is sought in the form

$$\psi = \psi_s(t) + \frac{z}{A(t)}, \tag{3}$$

where ψ_s is the surface pressure head and A is an unknown function of time. Further, define the soil moisture capacity as

$$C(t)\frac{d\theta}{d\psi} = \frac{dK}{d\psi}\frac{\frac{dA}{dt}}{\frac{d}{dt}[A(\psi_s - \psi)]} \tag{4}$$

where C is an arbitrary function of time. Note that ψ and t are independent variables in (4). With (4), substitution of (3) into (2) yields

$$A\frac{dA}{dt} = (1-A)\,C(t) \tag{5}$$

which has the solution

$$0 = A(t) - A(0) + ln\left[\frac{1-A(t)}{1-A(0)}\right] + \int_0^t C(\bar{t})\,d\bar{t}. \tag{6}$$

We consider soils which have a well defined air-entry value, given by ψ_a. Then, using (4), C is defined as

$$C(t) = \frac{1}{\Delta\theta}\int_{-\infty}^{\psi_a} \frac{\frac{dK}{d\psi}\frac{dA}{dt}}{\frac{d}{dt}[A(\psi_s - \psi)]}\,d\psi. \tag{7}$$

A slightly different definition of C is appropriate if the surface pressure head becomes lower than ψ_a. We focus here on infiltration under ponded conditions ($\psi_s > 0$), in which case (7) should be used (Barry et al., 1993b). With this definition, the soil moisture characteristic curve is

$$\frac{\theta - \theta_r}{\Delta\theta} = \begin{cases} \dfrac{\displaystyle\int_{-\infty}^{\psi} \frac{\frac{dK}{d\bar{\psi}}}{\frac{d}{dt}[A(\psi_s - \bar{\psi})]}\,d\bar{\psi}}{\displaystyle\int_{-\infty}^{\psi_a} \frac{\frac{dK}{d\psi}}{\frac{d}{dt}[A(\psi_s - \psi)]}\,d\psi}, & \psi < \psi_a. \\ 1, & \psi \geq \psi_a \end{cases} \tag{8}$$

This theoretical development is similar to that presented by Barry et al. (1993b), except that here the surface head is permitted to be a function of time. The right side of (8) is unrealistic because of the presence of the time-dependent surface head, whereas the soil moisture characteristic curve is usually taken to be independent of this quantities. However, it will be shown below that useful results can be obtained, particularly for soils which are of the Green and Ampt (1911) type. Before proceeding to the Green-Ampt case, we determine when the soil moisture characteristic curve is independent of time. Any surface pressure head that satisfies

$$\psi_s(t) = c_1 + c_2/A(t) \tag{9}$$

will reduce (8) to

$$\frac{\theta-\theta_r}{\Delta\theta} = \begin{cases} \dfrac{\displaystyle\int_{-\infty}^{\psi} \frac{\frac{dK}{d\bar{\psi}}}{c_1-\bar{\psi}}\,d\bar{\psi}}{\displaystyle\int_{-\infty}^{\psi_a} \frac{\frac{dK}{d\psi}}{c_1-\psi}\,d\psi}, & \psi < \psi_a. \\ 1, & \psi \geq \psi_a \end{cases} \tag{10}$$

The solution presented by Barry et al. (1993b) is for soil moisture characteristic curves defined by (10), with c_1 representing the (constant) surface pressure head. Further, (7) then shows that C is constant, and so (6) identical to (11) of Barry et al. (1993b). Thus, their results appear to apply not only to a fixed head surface pressure head, but also to the time dependent surface head given by (9). However, for this latter case, Barry et al. (1993b) showed that the solution (3) reduces to the case of a constant pressure head condition at the surface under a translation of distance c_2 in the profile.

Regardless of the functional form of $\psi_s(t)$, integration of Darcy's law gives the cumulative infiltration at the soil surface as

$$I = K_s\int_0^t 1 - \frac{1}{A(\bar{t})}\,d\bar{t}, \tag{11}$$

assuming that $\psi_s > \psi_a$.

3.1. Green-Ampt Soil

A Green-Ampt soil is one which has a hydraulic conduction function defined by

$$K(\psi) = K_s H(\psi - \psi_a), \tag{12}$$

where H is the Heaviside step function. Using (12), C is found from (7) to be

$$\frac{dA}{dt}C^{-1} = \frac{\Delta\theta}{K_s}\frac{d}{dt}[A(\psi_s - \psi_a)]. \tag{13}$$

Then, from (5), A is defined by the solution of

$$\frac{\Delta\theta}{K_s}\frac{d}{dt}[A(\psi_s - \psi_a)]A = 1 - A, \tag{14}$$

with $A(0) = 0$ (Barry et al., 1993b). For a Green-Ampt soil, (11) becomes

$$I = -\Delta\theta\, A(t)\,(\psi_s - \psi_a). \tag{15}$$

Now, using (15), (14) can be written as the familiar Green-Ampt infiltration equation

$$I\frac{dI}{dt} = K_s\Delta\theta(\psi_s - \psi_a) + K_s I. \tag{16}$$

Thus, the theory has reduced to the Green-Ampt model as it should. Exact infiltration formulas can be derived from (16) upon specification of suitable functional forms of ψ_s. For example, if a fixed head of water exists at the soil surface at $t = 0$ then, if the fluid is not replenished, the falling head Green-Ampt model results. Equation (16) is solved by setting $\psi_s = \psi_s(0) - I$. Since $I(0) = 0$, the solution to (16) can be written as (Philip, 1992a)

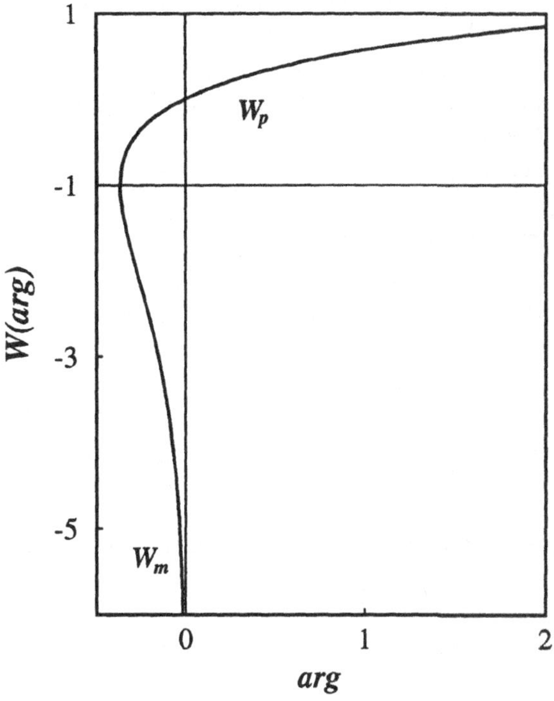

Figure 1. The function $W(arg)$. Above and below the line $W = -1$ lie the two branches W used here, i.e., W_p and W_m, respectively.

$$(1-\Delta\theta)^2 tK_s = I(1-\Delta\theta) - \Delta\psi\Delta\theta ln\left[1+\frac{I(1-\Delta\theta)}{\Delta\psi\Delta\theta}\right], \tag{17}$$

which can be expressed explicitly in the equivalent form

$$\frac{I(1-\Delta\theta)}{\Delta\psi\Delta\theta} = -1 - W_m\{-exp\left[-1-\frac{tK_s(1-\Delta\theta)^2}{\Delta\psi\Delta\theta}\right]\}. \tag{18}$$

The function $W(arg)$ is defined by solutions of (Fritsch et al., 1974)

$$W\,exp(W) = arg. \tag{19}$$

The behaviour of the function W is displayed in Fig. 1. Note that Barry et al. (1993b) give simple approximations for portion of this function in the region $arg < 1$. Barry et al. (1993a) have derived an efficient scheme to compute W for all arg. Obviously, other solutions can be determined for (16) upon specification of ψ_s (e.g., Philip, 1992b). However, we wish to consider more general soils than those defined by (12).

3.2. Solution for a General Soil

We consider soils having the property that their soil moisture characteristic curve satisfies (8). As mentioned already, (8) is not considered as a realistic model. Rather, it defines a class of soils for which exact results can be derived. Subsequently, these results will be used to test a general infiltration formula.

Equation (11) gives an exact infiltration formula in terms of the unknown function A, which is defined in turn by (5). A short time expansion of (5) in powers of $t^{1/2}$ can easily be calculated. It is not difficult to show as $t \to 0$ that

$$A = -\sqrt{\frac{2t}{\Delta\theta}\int_{-\infty}^{\psi_a}\frac{\frac{dK}{d\psi}}{\psi_s(0)-\psi}d\psi} + O(t). \tag{20}$$

Thus, from (11), the sorptivity for this class of soils is

$$S = K_s\sqrt{\frac{2\Delta\theta}{\int_{-\infty}^{\psi_a}\frac{\frac{dK}{d\psi}}{\psi_s(0)-\psi}d\psi}}. \tag{21}$$

An approximate solution for $A(t)$ can now be developed. First, we define X as

$$X(t) = \psi_s + A\frac{\frac{d\psi_s}{dt}}{\frac{dA}{dt}} = \psi_s + \frac{A^2\frac{d\psi_s}{dt}}{C(1-A)}. \tag{22}$$

With $A(0) = 0$, the solution to (5) is, in terms of the W function,

$$A = 1 + W_m\left\{-exp\left[-1-\int_0^t C(\bar{t})\,d\bar{t}\right]\right\}. \tag{23}$$

For a given K, (7) and (22) define C implicitly. In (23), however, the integral of C is needed. This can be achieved using a simple numerical approximation. Upon setting $t = n\Delta t$, and using the trapezoidal rule (Davis and Polonsky, 1964), (23) becomes

$$A(n\Delta t) = 1 + W_m\left\{-exp\left[-1-\frac{\Delta t}{2}C(0)-\frac{\Delta t}{2}C(n\Delta t)-\Delta t\sum_{i=1}^{n-1}C(i\Delta t)\right]\right\}. \tag{24}$$

Thus, (7) and (22) give a single equation with $A(n\Delta t)$ and $C(n\Delta t)$ as the unknowns, as does (24). However, (24) gives $A(n\Delta t)$ explicitly, so that the combination of (7) and (22) reduces to a single nonlinear equation with $C(n\Delta t)$ as the only unknown. These equations are solved stepwise in time starting at Δt using (20) to give $A(\Delta t)$. It is easily shown that, to the same order of accuracy, $C(\Delta t) = A(\Delta t)^2/2\Delta t$. With C, and hence A known, the cumulative infiltration can be determined from (11) using a convenient quadrature formula, for example, the trapezoidal rule. Accuracy is specified by suitable choice of Δt. The results, however, are limited to the class of soils defined by (8). One use of the exact solution is to check more general analytical approximations for infiltration. A general infiltration formula for a time-dependent head condition has been proposed (Parlange et al., 1985). The exact results presented here provide the opportunity to check directly the accuracy of this formula.

Before proceeding, we note that when the surface pressure head is constant, an infiltration equation of the form of (16) can always be derived following the approach used here, as shown by Barry et al. (1993b). For a time-dependent surface pressure head, such a simple formula is possible only in the case of the step function hydraulic conductivity, (12). Thus, the Green-Ampt assumption is extremely useful in simplifying the governing equation.

3.3. Infiltration Formula

Parlange et al. (1985) proposed an approximate infiltration formula for an arbitrary ponded, time-dependent surface condition, extending the earlier, less general, infiltration formula of Parlange et al. (1982). For infiltration into a dry soil, the Parlange et al. (1985) formula is:

$$I-\frac{K_s\psi_s(t)\Delta\theta(1+\mu)}{\frac{dI}{dt}-K_s}=\frac{S_0^2}{2\delta K_s}ln\left(1+\frac{\delta K_s}{\frac{dI}{dt}-K_s}\right), \tag{25}$$

where S_0 is the sorptivity for a zero surface head, and μ and δ are defined, respectively, by

$$[S_0^2+2K_s\psi_s(0)\Delta\theta]\mu=\int_{-\infty}^{\psi_a}[\theta_s-\theta(\psi)]K(\psi)\,d\psi \tag{26}$$

and

$$\delta=1-\frac{1}{\Delta\theta K_s}\int_{-\infty}^{\psi_a}\frac{d\theta}{d\psi}K(\psi)\,d\psi. \tag{27}$$

The parameters μ and δ are constants for any particular soil, and so (25) is a first-order differential equation for I which is solved numerically. The most convenient solution method results from rewriting (25) as

$$\frac{dI}{dt}-K_s=\frac{\delta K_s}{\frac{B}{H}-1}, \tag{28}$$

where

$$B=W_p\left[H exp\left(H+\frac{2\delta K_s I}{S_0^2}\right)\right] \tag{29}$$

and

$$H=\frac{2\psi_s(t)K_s\Delta\theta(1+\mu)}{S_0^2}. \tag{30}$$

The advantage of (28) is that it isolates the derivative term. Standard integration methods, such as the Runge-Kutta technique (e.g., Davis and Polonsky, 1964), can then be employed to solve (28) very easily and efficiently. The only difficulty with numerical integration of (28) is that $dI/dt \to \infty$ as $t \to 0$. However, in this limit, I is given by $St^{1/2}$. In general, S will be unknown, although it is assumed that S_0 is known. Then, S can be approximated by (Green and Ampt, 1911)

$$S^2=S_0^2+2K_s\psi_s(0)\Delta\theta. \tag{31}$$

Clearly, the approximation (31) has been used to derive (26). More accurate sorptivity approximations are available and have been extensively tested (Parlange, 1975; Barry et al., 1992; Parlange et al.,

1992, 1993), should they be required.

For a constant ψ_s, (28) can be integrated explicitly. The solution is

$$\frac{2t\delta K_s^2(1-\delta)}{S_0^2} = \delta ln\,(1+\frac{B-H}{\delta H})\,(H\delta - H - 1) + ln\,(\frac{B}{H}) - (H-B)\,(1-\delta). \tag{32}$$

Equation (32) provides a straightforward formula to calculate $I(t)$ for constant head infiltration.

3.3.1. Application to falling head infiltration. A simple calculation shows that (25) reduces to (16) in the limit $\delta \to 0$ and $\mu \to 0$ (Parlange et al., 1985). In other words, (25) is exact for the case of a Green-Ampt soil. As shown by (12), such soils are characterised by rapid changes in hydraulic conductivity around the air-entry value. To test (25) then, we choose a conductivity that is slowly varying, e.g.,

$$K = K_s\left(\frac{\psi_a}{\psi}\right)^2, \psi < \psi_a. \tag{33}$$

There is nothing particularly significant about this hydraulic conductivity, other than its slow variation. Indeed, it is unrealistic since the conductivity in most soils would vary much more rapidly than predicted by (33) (Reichardt et al., 1972).

Before solving (28), μ and δ must be calculated. Upon substituting (8) and (33), as necessary, into (26) and (27), we find, respectively,

$$\frac{-[S_0^2 + 2K_s\psi_s(0)\Delta\theta]\,\mu}{2K_s\psi_a\Delta\theta} = \frac{\psi_a^3 ln\,(1-\frac{X}{\psi_a}) + \psi_a^2 X + \frac{X^2\psi_a}{2} + \frac{X^3}{3}}{X\left[2\psi_a X + X^2 + 2\psi_a^2 ln\,(1-\frac{X}{\psi_a})\right]} \tag{34}$$

and

$$\frac{(1-\delta)X^2}{2} = \frac{\psi_a^4 ln\,(1-\frac{X}{\psi_a}) + X\left(\frac{\psi_a X^2}{3} + \frac{X^3}{4} + \frac{X\psi_a^2}{2} + \psi_a^3\right)}{2\psi_a X + X^2 + 2\psi_a^2 ln\,(1-\frac{X}{\psi_a})}. \tag{35}$$

Equations (34) and (35) are functions of time. This is due to the time-dependence of the soil moisture characteristic curve, (8). Of course, for real soils, both μ and δ are constants. Also, the functional form of X means that (28) is no longer an explicit function of dI/dt, since (11) can be used to eliminate dA/dt and A from X. We found, however that μ and δ are only slowly varying functions of time, and can be replaced by average values. This point will be addressed below. However, we note that using average values for these parameters is clearly a more stringent test of (28).

For the purpose of illustration, we used the following parameters to carry out the calculations: $\psi_s(0) = 10$ cm, $\psi_a = -1$ cm, $K_s = 1$ cm d^{-1} and $\Delta\theta = 0.5$. With these values we first computed the exact cumulative infiltration using (7), (11), and (24). The exact results were used to compute μ and δ from (34) and (35). The behaviour of these parameters, along with that of X, is shown in Fig. 2. Note that in this example the surface ceases to be ponded at $t \approx 5.56$ d. Figure 2a shows that the μ varies only marginally, while the variation of δ is more marked (Fig. 2b). The variation of X is in the range $\psi_s(0) \geq X > \psi_a$ (Fig. 2c). A suitable average X for the calculation of the average μ and δ is clearly X(2.28 d). However, there is no reason that this time would be known *a priori*. A more reasonable choice is to calculate μ and δ using a fraction of $\psi_s(0)$, rather than X. For the sake of simplicity we selected $\psi_s(0)/2 = 5$ cm. Equations (34) and (35) then give $\mu(5\text{ cm}) = 0.03$ and $\delta(5\text{ cm}) = 0.47$. As can be seen from Fig. 2b, this

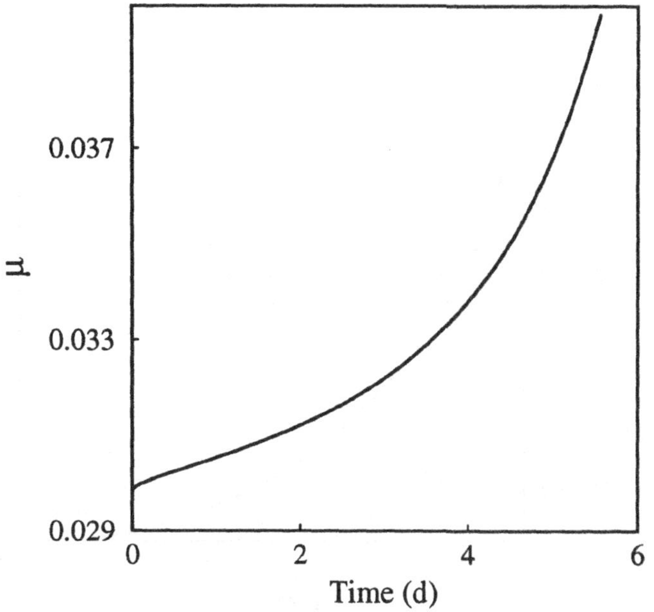

Figure 2a. Variation of μ with time for the falling head infiltration example.

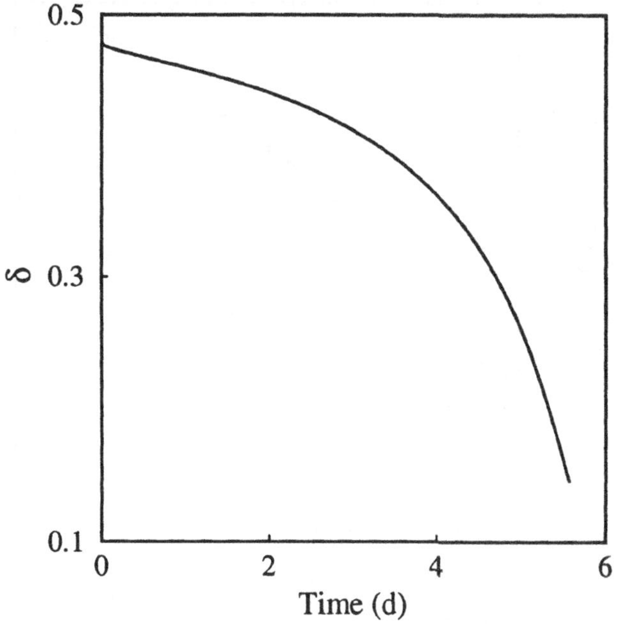

Figure 2b. Variation of δ with time for the falling head infiltration example.

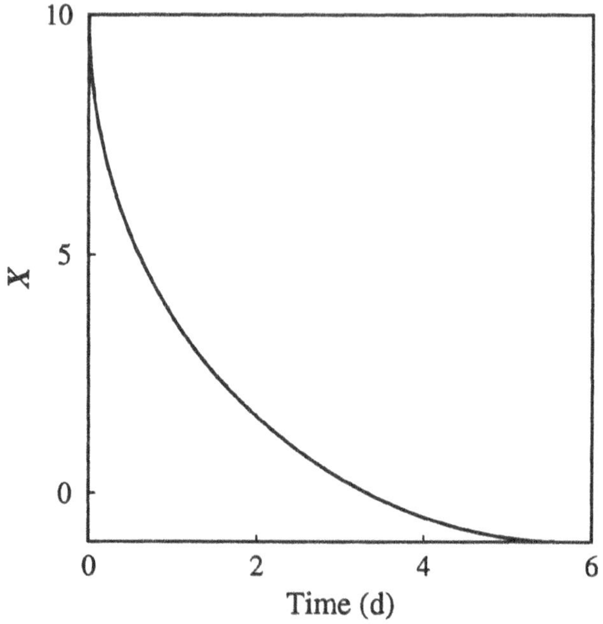

Figure 2c. Variation of X with time for the falling head infiltration example.

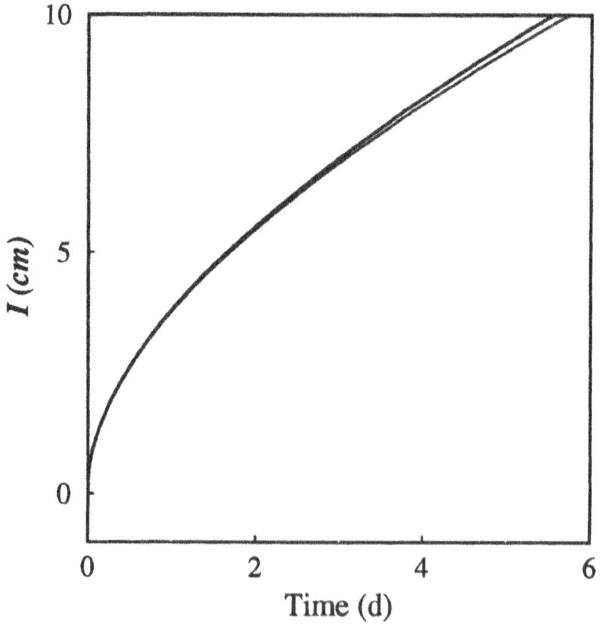

Figure 3. Comparison of exact (thick, upper) and approximate (thin, lower) cumulative infiltration curves for falling head infiltration.

is not a particularly judicious choice for δ, as it corresponds to $t \approx 0.6$ d.

Equation (28) was solved numerically using the Runge-Kutta method, with the parameter values discussed above. The prediction from (28), and the exact falling head solution (calculated following the discussion in section 3.2), is plotted for comparison in Fig. 3. The predicted infiltration is slightly less than the exact value, with a maximum relative error of approximately 2% occurring at the conclusion of the infiltration event. Given the artificial nature of the 'soil' used in this comparison, this result suggests that (28) is a robust approximation.

3.3.2. Application to a time-dependent surface condition. As a second example, we take the following condition:

$$\psi_s = \psi_a + \gamma + \beta exp(-\varepsilon t) \tag{36}$$

with the same conductivity function as above. In addition, γ and β were taken as 6 and 5 cm, respectively, with $\varepsilon = 1\ d^{-1}$. Again, there is nothing particularly significant about this surface condition. As above, it was found from the exact solution that μ and δ varied little. Thus, X was approximated by ψ_s. The calculations were performed as already described. The cumulative infiltration is shown in Fig. 4. The error, like that in Fig. 3, grows with time at a small rate. The maximum relative error in the figure is less than 2%.

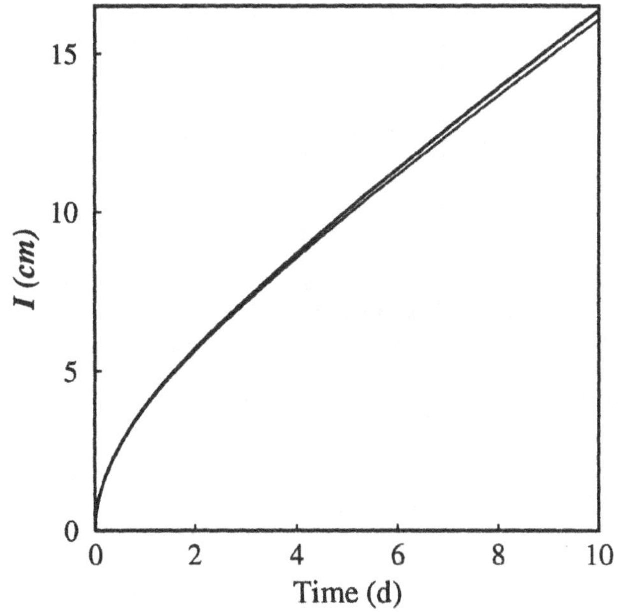

Figure 4. Comparison of exact (thick, upper) and approximate (thin, lower) cumulative infiltration curves for the time-dependent surface condition (36).

3.4 Drainage from a Soil Profile

We consider the saturated-unsaturated soil profile shown in Fig. 5. Initially, the external reservoir, which controls the head at the profile base, is located such that the air-water interface is z_I above the base. The z-axis has its origin at the profile and, as above, is oriented so that positive z is downwards. Within the soil profile, the soil is saturated from $z = 0$ up to $z = \psi_a - z_I$. Dagan and Kroszynski (1973) and Dagan (1989) consider the drainage that will occur from the soil profile if the external reservoir is suddenly dropped so that the phreatic surface in the reservoir is aligned with the bottom of the profile, i.e., $z = 0$.

The drainage of the column is governed by Richards' equation (1). The initial time, $t = 0$, is taken as the instant the external reservoir is lowered. Then, the initial pressure head gradient is

$$\frac{\partial\psi}{\partial z} = 1 - \frac{z_I}{z_I - \psi_a}, \qquad t = 0. \tag{37}$$

The solution to (1) is given by (3) with $\psi_s = 0$. Equations (3) and (37) show that

$$A(0) = 1 - \frac{z_I}{\psi_a}. \tag{38}$$

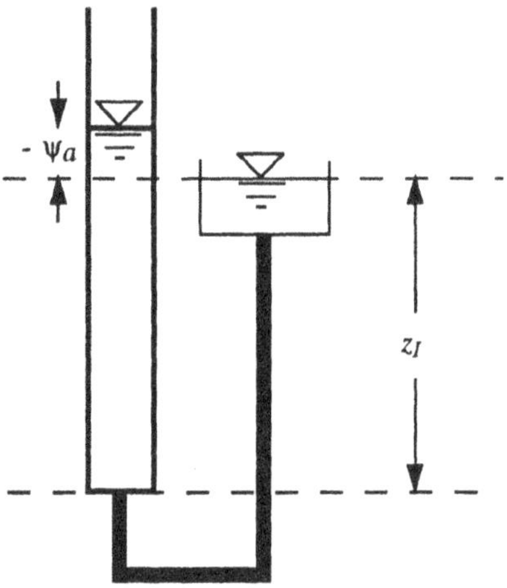

Figure 5. Initial configuration of the drainage problem: Due to the external fluid reservoir, the soil profile is saturated as shown to depth $z_I - \psi_a$. At $t = 0$, the external reservoir is lowered so that the phreatic surface of the reservoir fluid is aligned with the base of the soil profile.

Equation (6) is the solution for $A(t)$. Because ψ_s is constant, $C(t)$, which is calculated from (7), is a constant independent of t. Denote this constant value by C_c. Now, the relationship between the cumulative drainage, D_r, and A is just (11), with I replaced by D_r. Integration of (11) yields

$$D_r = \Delta\theta\psi_a[A(t) - A(0)]. \tag{39}$$

With (39) and (6), D_r can be expressed as

$$\frac{D_r}{\Delta\theta\psi_a} + ln\left(1 - \frac{D_r}{\Delta\theta z_I}\right) + C_c t = 0. \tag{40}$$

An explicit solution for D_r is available using the W function:

$$\frac{D_r}{\Delta\theta z_I} = 1 + \frac{\psi_a}{z_I} W_p\left[-\frac{z_I}{\psi_a} exp\left(-\frac{z_I}{\psi_a} - C_c t\right)\right]. \tag{41}$$

Dagan (1989) derived an expression which, under suitable notation changes, is identical to (40). The predictions of the formula were compared with the extensive experimental data set of Vachaud (1968) with reasonably good results. However, to derive his result, Dagan (1989) assumed that the capillary fringe remains fixed during the entire drainage process. The derivation given here relies not on this assumption, but rather on the relationship (8). Clearly, the results could be extended to account for a time-dependent bottom boundary condition, along the lines of the analysis of the previous sections.

3.5 Application to Solute Transport

Barry et al. (1991) showed that if K is defined as (Gardner, 1958)

$$K = \frac{v}{B} exp\left(\frac{v\psi}{D}\right), \tag{42}$$

then with the mapping

$$K = \frac{vc}{B} \tag{43}$$

and

$$\theta = \frac{c + s(c)}{B}, \tag{44}$$

(1) becomes

$$\frac{\partial[c + s(c)]}{\partial t} = D\frac{\partial^2 c}{\partial z^2} - v\frac{\partial c}{\partial z}, \tag{45}$$

which is the governing equation for solute transport with the nonlinear sorption isotherm, s (e.g., Bajracharya and Barry, 1993). Barry et al. (1993b) have presented a new solution for (45) based on (3) for ψ_s constant. Here, this solution is extended to the case of a time-dependent ψ_s.

The soil profile is considered to be initially solute free, i.e.

$$c(z,0) = 0, \qquad z > 0, \tag{46}$$

while at the profile surface the solute concentration is a known function of time:

$$c(0,t) = c_0(t), \qquad t > 0. \tag{47}$$

The solution to (45)-(47) is, from (3) and (43)

$$c = \begin{cases} c_B exp\left(\frac{vz}{DA}\right), z > z_f \\ c_0, z < z_f \end{cases}, \tag{48}$$

where $c_B(t)$ is an isotherm fitting parameter defined by

$$c_B = exp\left(\frac{v\psi_s}{D}\right) \tag{49}$$

and z_f is the position in the profile locating the solute front position where $c = c_0$,

$$z_f = \frac{AD}{v} ln\left(\frac{c_0}{c_B}\right). \tag{50}$$

The surface concentration, c_0, is related to the air-entry pressure head via

$$c_0 = exp\left(\frac{v\psi_a}{D}\right). \tag{51}$$

Because c_0 is a time dependent, so to is ψ_a in (51). We note that this presents no difficulties in obtaining the solution from (43) as this time dependence enters in the calculation of $C(t)$, as given by (7). Equations (48)-(51) relate various quantities needed in the solution for (45)-(47) to the solution already obtained for Richards' equation (1). The isotherm, s, for which (48) is an exact solution is then

$$\frac{s(c)+c}{s(c_0)+c_0} = \frac{Ei\left[ln\left(\frac{c}{c_B}\right) - \frac{A\frac{dc_B}{dt}}{c_B\frac{dA}{dt}}\right]}{Ei\left[ln\left(\frac{c_0}{c_B}\right) - \frac{A\frac{dc_B}{dt}}{c_B\frac{dA}{dt}}\right]}, \tag{52}$$

where Ei is the exponential integral (Spanier and Oldham, 1987). Note that, to derive (52), we used the relationships:

$$\theta(c) = -Ei\left[ln\left(\frac{c}{c_B}\right) - \frac{A\frac{dc_B}{dt}}{c_B\frac{dA}{dt}}\right], \tag{53}$$

and

$$B = \frac{-[s(c_0)+c_0]}{Ei\left[ln\left(\frac{c_0}{c_B}\right) - \frac{A\frac{dc_B}{dt}}{c_B\frac{dA}{dt}}\right]}. \tag{54}$$

Since c_0 is defined by the boundary condition (47), the isotherm given by (52) has c_B as a fitting parameter, as already noted. Barry et al. (1993b) have shown that this isotherm, for c_B constant, is similar to the lower portion of the S-curve isotherm (Sposito, 1989; Barry, 1992). Such isotherms occur, e.g., in cation exchange processes (Schweich et al., 1983). In addition, Barrow (1989) has pointed out that time-dependence of solute sorption in soils is much more common than equilibrium sorption. Thus, it appears that the time dependence of the adsorption isotherm, (52), allows for more realistic modelling of solute sorption behaviour. Finally, the total amount of solute, I_c, that has entered the soil at any time is found by integrating the solution (48) over the spatial domain. The result is

$$I_c = c_0\left(z_f - \frac{AD}{v}\right). \tag{55}$$

4. Conclusions

There are very few exact solutions for Richards' equation, the governing equation for unsaturated water flow. Recently, some recent exact solutions have been derived using specific forms of the diffusivity and conductivity functions (Rogers et al., 1983; Broadbridge and White, 1988; Sander et al., 1988; Barry and Sander, 1991; Warrick et al., 1990, 1991). The solution presented here, and that of Barry et al. (1993b), do not depend on specific soil hydraulic functions, but on a relationship between the soil moisture characteristic curve, the hydraulic conductivity, and the surface pressure head. Of course, this relationship is constraining, and the soil to which the solution applies is, consequently, of an artificial nature. On the other hand, the solution obtained is relatively simple. Hence, it should be of use in checking numerical schemes, where no other checks are possible due to the lack of analytical solutions for nonlinear forms of Richards' equation. For the draining soil profile, a particularly simple drainage formula can be derived. Finally, the solution for infiltration has been used to derive an exact solution for nonlinear sorption of a solute moving in a soil profile. The sorption isotherm depends on time through the surface condition, as well as through a fitting parameter. Again, this solution is one of a very few exact solutions for nonlinear solute adsorption and transport.

References

Bajracharya, K., and D. A. Barry. 'Mixing cell models for nonlinear equilibrium single species adsorption and transport'. *J. Contam. Hydrol.* **12**:227-243.

Barrow, N. J. 1989. 'Suitability of sorption-desorption models to simulate partitioning and movement of ions in soils'. In B. Bar-Yosef, N. J. Barrow, and J. Goldsmith (eds.), *Inorganic Contaminants in the Vadose Zone*. Springer-Verlag, Berlin.

Barry, D. A. 1992. 'Modeling contaminant transport in the subsurface: Theory and computer programs'. In H. Ghadiri, and C. W. Rose (eds.), *Modeling Chemical Transport in Soils, Natural and Applied Contaminants*, Lewis Publishers, Boca Raton, Florida.

Barry, D. A., S. J. Barry, and P. J. Hensley. 1993a. 'Real values of the W function'. Submitted to *ACM Trans. Math. Software.*

Barry, D. A., J.-Y. Parlange, R. Haverkamp, and J. Touma. 1992. 'Comment on "Solution of a nonlinear absorption model of mixed saturated-unsaturated flow" by P. Broadbridge'. *Water Resour. Res.* **28**:1737-1738.

Barry, D. A., J.-Y. Parlange, G. C. Sander, and M. Sivapalan. 1993b. 'A class of exact solutions for Richards' equation'. *J. Hydrol.* **142**:29-46.

Barry, D. A., and G. C. Sander. 1991. 'Exact solutions for water infiltration with an arbitrary surface flux or nonlinear solute adsorption'. *Water Resour. Res.* **27**:2667-2680.

Barry, D. A., G. C. Sander, and I. R. Phillips. 1991. 'Modelling of solute transport, chemical adsorption and cation exchange'. *Int. Hydrology and Water Resources Symp.* 2-4 October, Perth, W.A. The Institution of Engineers, Australia, Nat. Conf. Publ. No. 92/19 (Preprints of Papers, Vol. 3:924-929).

Broadbridge, P., and I. White. 1988. 'Constant rate rainfall infiltration: A versatile nonlinear model, 1, Analytic solution'. *Water Resour. Res.* **24**:145-154.

Dagan, G. 1989. *Flow and Transport in Porous Formations*. Springer-Verlag, Berlin.

Dagan, G., and U. Kroszynski. 1973. 'Drainage of a vertical column'. In A. Hadas, D. Swartzendruber, P. E. Rijtema, M. Fuchs, and B. Yaron (eds.), *Physical Aspects of Soil Water and Salts in Ecosystems.* Chapman and Hall Limited, London.

Davis, P. J., and I. Polonsky. 1964. 'Numerical interpolation, differentiation and integration'. In M. Abramowitz and I. A. Stegun (eds.), *Handbook of Mathematical Functions With Formulas, Graphs, and Mathematical Tables*. National Bureau of Standards, Appl. Math. Ser. 55, Washington, D.C.

Fritsch, F. N., R. E. Shafer, and R. E. Crowley. 1974. 'Solution of the transcendental equation $we^w = x$'. *Commun. Assoc. Comput. Mach.* **16**:123-124.

Gardner, W. R. 1958. 'Some steady state solutions of the unsaturated moisture flow equation with application to evaporation from a water table'. *Soil Sci.* **85**:228-232.

Green, W. H., and G. A. Ampt. 1911. 'Studies in soil physics: 1. The flow of air and water through soils'. *J. Agric. Sci.* **4**:1-24.

Onstad, C. A., T. C. Olson, and L. R. Stone. 1973. 'An infiltration model tested with monolith moisture measurements'. *Soil Sci.* **116**:13-17.

Parlange, J.-Y. 1975. 'On solving the flow equation in unsaturated soils by optimization: Horizontal infiltration'. *Soil Sci. Soc. Am. Proc.* **39**:415-418.

Parlange, J.-Y., D. A. Barry, M. B. Parlange, and R. Haverkamp. 1992. 'Note on the sorptivity for mixed saturated-unsaturated flow'. *Water Resour. Res.* **28**:2529-2531.

Parlange, J.-Y., D. A. Barry, M. B. Parlange, D. A. Lockington, and R. Haverkamp. 1993. 'Sorptivity calculation for arbitrary diffusivity'. *Transp. Porous Media* In Press.

Parlange, J.-Y., R. Haverkamp, and J. Touma. 1985. 'Infiltration under ponded conditions: 1. Optimal analytical solution and comparison with experimental observations'. *Soil Sci.* **139**:305-311.

Parlange, J.-Y., I. Lisle, R. D. Braddock, and R. E. Smith. 1982. 'The three-parameter infiltration equation'. *Soil Sci.* **133**:337-341.

Philip, J. R. 1992a. 'Falling head ponded infiltration'. *Water Resour. Res.* **28**:2147-2148.

Philip, J. R. 1992b. 'Falling head ponded infiltration with evaporation'. *J. Hydrol.* **138**:591-598.

Rogers, C., M. P. Stallybrass, and D. L. Clements. 1983. 'On two phase filtration under gravity and with boundary infiltration: Application of a Bäcklund transformation'. *Nonlin. Anal. Theory Meth. Appl.* **7**:785-799.

Reichardt, K., D. R. Neilsen, and J. W. Biggar. 1972. 'Scaling of horizontal infiltration into homogeneous soils'. *Soil Sci. Soc. Am. J.* **36**:241-245.

Richards, L. A. 1931. 'Capillary conduction of liquids through mediums'. *Physics* **1**:318-333.

Sander, G. C., J.-Y. Parlange, V. Kühnel, W. L. Hogarth, D. Lockington, and J. P. J. O'Kane. 1988. 'Exact nonlinear solution for constant flux infiltration'. *J. Hydrol.* **97**:341-346.

Schweich, D., M. Sardin, and J. P. Gaudet. 1983. 'Measurement of a cation exchange isotherm from elution curves obtained in a soil column: Preliminary results'. *Soil Sci. Soc. Am. J.*, **47**:32-37.

Spanier, J., and K. B. Oldham. 1987. *An Atlas of Functions*. Hemisphere Publishing Corp., New York.

Sposito, G. 1989. *The Chemistry of Soils*. Oxford University Press, New York.

Vachaud, G. 1968. *Contribution a L'Etude des Problemes D'Ecoulement en Milieux Poreux Nonsaturees*. Ph.D. thesis, Grenoble, France.

Warrick, A. W., D. O. Lomen, and A. Islas. 1990. 'An analytical solution to Richards' equation for a draining soil profile'. *Water Resour. Res.* **26**:253-258.

Warrick, A. W., A. Islas, and D. O. Lomen. 1991. 'An analytical solution to Richards' equation for time-varying infiltration'. *Water Resour. Res.* **27**:763-766.

SOME NUMERICAL EXPERIMENTS ON THE VARIABLY–SATURATED FLOW EQUATION

Chittaranjan Ray

Illinois State Water Survey
Hydrology Division
2204 Griffith Drive
Champaign, IL 61820

(Also a doctoral student in Civil Engineering, University of Illinois at Urbana–Champaign)

Abstract

The variably–saturated flow equation (VSFE) is a special form of the well known Richards' equation for unsaturated flow in which the elastic storage arguments are used to define the specific storativity (which is useful for saturated flow simulation). The Richards' equation, under transient conditions, is highly nonlinear. Traditionally, numerical techniques have been used in solving this equation. The moisture–content form of Richards' equation requires less computer time and is subject to low mass balance errors while simulating transient infiltration into very dry soils. However, this form of Richards' equation is limited to complete unsaturated conditions and cannot handle positive pressure upper boundary or layered soil profiles. For this reason, while simulating water movement in variably saturated soils, the pressure–based form of Richards' equation is used. The pressure–based form is prone to mass balance errors and requires large amounts of computer time due to the iterative nature of solution. A recently reported mixed form of Richards' equation is used in this investigation for its mass conservative property. The computer time requirement for the mixed form Richards' equation is similar to the pressure–based form. A series of numerical experiments were conducted using the VSFE which contained the mass–conservative form of the Richards' equation. The Galerkin finite element technique was used to solve the equation numerically. The effects of the type of mass lumping (matrix diagonalization), initial and boundary conditions on the accuracy, stability, and the rate of convergence of this equation are presented. The influences of time varying top boundary conditions and layered heterogeneity on water movement are also presented. In addition, "tricks" for efficient time step control and steady–state simulation using this mass conservative formulation are shown. Finally, simulation results indicate that this mixed–form equation is superior to the pressure–based form.

V. P. Singh and B. Kumar (eds.), Subsurface-Water Hydrology, 49–63.

Introduction

Flow of water in the vadose zone plays an important role in predicting ground–water recharge and contaminant migration from land surface to water table. The role of the vadose zone in contaminant migration is receiving greater attention in recent years because of the complexities involved in estimating the flow parameters for the variably saturated flow equation (VSFE). A number of computer models have been developed in the past ten years to study the mechanisms of water flow and contaminant migration in unsaturated soils.

Water flow in the unsaturated soil is described by Richards' equation. In one–dimensional form, this equation is written as

$$\frac{\partial \theta}{\partial t} + \frac{\partial}{\partial z}\left[K - K\frac{\partial h}{\partial z}\right] = 0 \tag{1}$$

where θ is the water content (volumetric) [L^3/L^3], h is the pressure head [L], t is time [T], K is the hydraulic conductivity [L/T], and z is the coordinate direction, measured positive downward in equation (1). If the elastic storage arguments are used for the simulation of saturated and unsaturated flow (Bear, 1979), equation (1) becomes:

$$\frac{\partial \theta}{\partial t} + \left(S_s\frac{\theta}{\theta_s}\frac{\partial h}{\partial t}\right) + \frac{\partial}{\partial z}\left[K - K\frac{\partial h}{\partial z}\right] = 0 \tag{2}$$

where S_s is the elastic storage coefficient [1/L] and θ_s is the saturated water content. The ratio, θ/θ_s is the index of saturation, which varies between zero and one. As can be observed, equations (1) and (2) contain two unknowns, the moisture content (θ) and the pressure head (h). It is customary to write (1) in terms of the pressure head

$$C\frac{\partial h}{\partial t} + \frac{\partial}{\partial z}\left[K - K\frac{\partial h}{\partial z}\right] = 0 \tag{3}$$

where the coefficient C is defined as the specific moisture capacity [1/L]. This is expressed as

$$C \equiv \frac{d\theta}{dh} \tag{4}$$

which is the slope of the moisture content–pressure head graph. The corresponding form of pressure–based equation with elastic storage expression is given as:

$$C\frac{\partial h}{\partial t} + \left(S_s\frac{\theta}{\theta_s}\frac{\partial h}{\partial t}\right) + \frac{\partial}{\partial z}\left[K - K\frac{\partial h}{\partial z}\right] = 0 \tag{5}$$

In equations (2) and (4), both K and C are highly nonlinear functions of pressure head. Numerical methods, e.g. finite element or finite difference, are commonly employed to solve equations (2) and (4) with appropriate initial and boundary conditions. This paper examines the numerical analysis aspects of solving equation (2).

History of Modeling Efforts

A majority of the computer models developed in the past decade use the pressure–based form of Richards' equation (3) to describe fluid flow in partially saturated soil. For example, the model UNSAT2 (Davis and Neuman, 1983) was developed to simulate fluid flow in multidimensional variably saturated media with ability to simulate extraction of water by plants and pumping wells. The model SATURN (Huyakorn et al., 1983) was developed to simulate fluid flow and radionuclide transport in one–space dimension. VAM2D is a comprehensive, two–dimensional, water flow and solute transport model (Huyakorn et al., 1989) which considers hysteresis for water flow and chained decay processes for chemical transport. In addition, the model HYDRUS (Kool and van Genuchten, 1991) was developed by the U.S. Salinity Laboratory of the U.S. department of Agriculture to simulate one–dimensional fluid flow and contaminant transport. The flow model takes into account the soil water hysteresis and the uptake of water by the plant roots. The transport model in HYDRUS is based upon equilibrium controlled chemical adsorption and first–order decay processes. Most of the above models (UNSAT2, SATURN, and VAM2D) are capable of simulating both unsaturated and saturated flow. They all use Galerkin finite element methods for spatial discretization of the domain and the pressure–based form of Richards' equation (5).

Because of the highly nonlinear nature of K and C in equations (2) and (5), most numerical schemes use some kind of iterative technics to solve these equations. Most of the models described above produce large mass balance errors while simulating water infiltration into very dry soils. The measure of this inaccuracy is expressed by the term "mass balance error" (MBR). The MBR is the ratio of the amount of mass (water or contaminant) added to the system and the amount of mass present in the system after a given elapsed time. A mass conservative system produces a MBR equal to 1. Milly (1985) proposed a new procedure to evaluate the specific moisture capacity (C) term in the element matrices to reduce mass balance errors. The model HYDRUS incorporates this technique for the reduction of mass balance errors in the flow segment of the code. In general, all pressure based models, require a significant amount of CPU time for realistic simulation events.

Prior to realizing the numerical difficulty of solving the pressure–based form of Richards' equation, soil scientists have expressed this equation in terms of the moisture content

$$\frac{\partial \theta}{\partial t} + \frac{\partial}{\partial z}\left[K - \frac{K}{C}\frac{\partial \theta}{\partial z}\right] = 0 \tag{6}$$

Childs and Collis–George (1950) originally defined the ratio K/C as the soil water diffusivity (D) $[L^2/T]$ and this form of Richards' equation is analogous to the heat conduction or the diffusion equation for which analytic solutions exist. Equation (6) contains the single unknown variable θ and the diffusivity function can be calculated from the above ratio. For numerical solution, it should be pointed out that the moisture content, in general, is a discontinuous function at the interface of two soil layers with contrasting properties (where as the pressure is continuous). Hills et al. (1989) proposed a finite difference model to solve equation (6) for layered soils. They considered the continuity of fluid flux at the layer interface and were able to simulate infiltration of water into very dry soils (up to −50,000 cm of suction). The CPU time requirements were one to three orders of magnitude lower than the equivalent pressure–based form. They also reported that the mass balance errors were low due to the noniterative nature of water–content formulation. Only round–off errors were present. Equation (6), however, cannot be applied to combined saturated–unsaturated or near saturated flow since the soil water capacity (C) approaches a value of zero near saturation. Local heterogeneity and layering often produce saturation areas in the domain and the numerical model of Hills et al. (1989) is not suitable for those cases. Further, this algorithm cannot handle a positive pressure upper boundary condition, especially associated with ponding after heavy precipitation or irrigation.

Celia et al. (1990) proposed a new numerical scheme to solve equation (1) directly. This method was shown by them to be completely mass conservative. To describe the procedure, a discretized form of equation (1) can be written using the finite element or the finite difference approach. In one–dimension, both methods produce identical spatial discretization if one assumes arithmetic averaging for conductivity between nodal values in the finite difference method. The only difference between the two methods is the time derivative terms, however, a globally mass lumped (diagonalized) finite element method is identical to the finite difference method. In this paper, the Galerkin finite element method is used for illustration purposes. Celia et al. (1990) compared both Picard and Newton–Raphson techniques to solve equation (1). The Picard iterative solution technique is conceptually simple and is utilized here for illustration purposes. In this method, the projected value at the new iteration level is guessed (usually as the initial value) and the unknown is solved. This iterative technique continues until the difference between the projected and the calculated values reaches below a preset tolerance. The time derivative part of equation (1) was discretized by Celia et al. (1990) using finite differences:

$$\frac{\theta_i^{n+1,m+1} - \theta_i^n}{\Delta t} - \frac{\partial}{\partial z}K^{n+1,m}\frac{\partial h^{n+1,m+1}}{\partial z} + \frac{\partial K^{n+1,m}}{\partial z} = 0 \tag{7}$$

where the superscripts n and n+1 refer to old and new time levels, respectively. Similarly, the superscripts m and m+1 refer to the current and the projected iteration values. They replaced the moisture content at the new time step and new iteration levels with the following Taylor series expansion

$$\theta_i^{n+1,m+1} = \theta_i^{n+1,m} + \frac{d\theta_i^{n+1,m}}{dh}\left(h_i^{n+1,m+1} - h_i^{n+1,m}\right) + \text{(higher order terms)} \tag{8}$$

After neglecting the higher order terms, the authors substituted equation (8) to (7). For convenience, they expressed $\delta h^{n+1,m} = h^{n+1,m+1} - h^{n+1,m}$, where the left hand side is treated as the unknown. The authors represented the unknowns δh, θ, K, and C by interpolating polynomials for the spatial discretization of equation (7) in the Galerkin finite elements. Subtraction of $h^{n+1,m}$ from each $h^{n+1,m+1}$, and the addition of the same terms to the right hand side to maintain algebraic equivalence, resulted in the following equation at each node:

$$\begin{aligned}
&\left[\left(C_{i-1}^{n+1,m} + C_i^{n+1,m}\right)\frac{(\Delta z)}{12\Delta t} - \frac{\left(K_{i-1}^{n+1,m} + K_i^{n+1,m}\right)}{2\Delta z}\right]\delta h_{i-1}^{n+1,m} + \left[\left(C_{i-1}^{n+1,m} + 6C_i^{n+1,m} + C_{i+1}^{n+1,m}\right)\frac{(\Delta z)}{12\Delta t} + \right.\\
&\left.\frac{\left(K_{i-1}^{n+1,m} + 2K_i^{n+1,m} + K_{i+1}^{n+1,m}\right)}{2\Delta z}\right]\delta h_i^{n+1,m} + \left[\left(C_i^{n+1,m} + C_{i+1}^{n+1,m}\right)\frac{(\Delta z)}{12\Delta t} - \frac{\left(K_i^{n+1,m} + K_{i+1}^{n+1,m}\right)}{2\Delta z}\right]\delta h_{i+1}^{n+1,m}\\
&= \left[\frac{\left(K_{i-1}^{n+1,m} + K_i^{n+1,m}\right)}{2\Delta z}\right]h_{i-1}^{n+1,m} - \left[\frac{\left(K_{i-1}^{n+1,m} + 2K_i^{n+1,m} + K_{i+1}^{n+1,m}\right)}{2\Delta z}\right]h_i^{n+1,m} + \left[\frac{\left(K_i^{n+1,m} + K_{i+1}^{n+1,m}\right)}{2\Delta z}\right]h_{i+1}^{n+1,m}\\
&- \left[\left(\theta_{i-1}^{n+1,m} + 4\theta_i^{n+1,m} + \theta_{i+1}^{n+1,m}\right)\frac{(\Delta Z)}{6\Delta t}\right] + \left[\left(\theta_{i-1}^n + 4\theta_i^n + \theta_{i+1}^n\right)\frac{(\Delta Z)}{6\Delta t}\right] + \frac{\left(K_{i-1}^{n+1,m} - K_{i+1}^{n+1,m}\right)}{(\Delta z)}
\end{aligned} \tag{9}$$

The system of algebraic equations produced by (9) can be written in the following matrix form

$$[A]\ \{\delta h^{n+1,m}\} = \{b\} \tag{10}$$

where $\delta h^{n+1,m}$ is defined earlier and [A] is a tridiagonal matrix. The right hand side vector {b} is a measure of the residual since we solve for δh.

The discretized form of equation (3) (Zarba, 1988) can be written as

$$\begin{aligned}
&\left[\left(C_{i-1}^{n+1,m} + C_i^{n+1,m}\right)\frac{(\Delta z)}{12\Delta t} - \frac{\left(K_{i-1}^{n+1,m} + K_i^{n+1,m}\right)}{2\Delta z}\right]\delta h_{i-1}^{n+1,m} + \left[\left(C_{i-1}^{n+1,m} + 6C_i^{n+1,m} + C_{i+1}^{n+1,m}\right)\frac{(\Delta z)}{12\Delta t} + \right.\\
&\left.\frac{\left(K_{i-1}^{n+1,m} + 2K_i^{n+1,m} + K_{i+1}^{n+1,m}\right)}{2\Delta z}\right]\delta h_i^{n+1,m} + \left[\left(C_i^{n+1,m} + C_{i+1}^{n+1,m}\right)\frac{(\Delta z)}{12\Delta t} - \frac{\left(K_i^{n+1,m} + K_{i+1}^{n+1,m}\right)}{2\Delta z}\right]\delta h_{i+1}^{n+1,m}\\
&= \left[-\left(C_{i-1}^{n+1,m} + C_i^{n+1,m}\right)\frac{(\Delta z)}{12\Delta t} + \frac{\left(K_{i-1}^{n+1,m} + K_i^{n+1,m}\right)}{2\Delta z}\right]h_{i-1}^{n+1,m} + \left[-\left(C_{i-1}^{n+1,m} + 6C_i^{n+1,m} + C_{i+1}^{n+1,m}\right)\frac{(\Delta z)}{12\Delta t}\right.\\
&\left.- \frac{\left(K_{i-1}^{n+1,m} + 2K_i^{n+1,m} + K_{i+1}^{n+1,m}\right)}{2\Delta z}\right]h_i^{n+1,m} + \left[-\left(C_i^{n+1,m} + C_{i+1}^{n+1,m}\right)\frac{(\Delta z)}{12\Delta t} + \frac{\left(K_i^{n+1,m} + K_{i+1}^{n+1,m}\right)}{2\Delta z}\right]h_{i+1}^{n+1,m}\\
&+ \frac{\left(K_{i-1}^{n+1,m} - K_{i+1}^{n+1,m}\right)}{(\Delta z)}
\end{aligned} \tag{11}$$

As evident from (9) and (11), both equations are almost identical except for the time derivative terms in the right hand side. As per Celia et al. (1990) this difference makes equation (9) mass conservative and termed this equation as the mass conservative (mixed form) "modified Picard" formulation. Since the left hand side of both (9) and (11) are same, the computational effort in both schemes is similar.

Recently, Kirkland et al. (1992) presented algorithms to solve the Richards' equation using two methods which require much lower CPU than the pressure–based algorithm. Using a new variable, they transform the Richards' equation (see Hills et al., 1989) which shows the characteristics of the water content when the soil is completely unsaturated and that of pressure at or near saturation. The second method is an improved pressure based method which uses a flux updating procedure for simulating flow from one layer to another. Both methods use noniterative techniques for solution. They suggested that this method introduces slight mass balance errors near saturation and that is acceptable from the CPU time requirement point of view. As reported by these authors, both the methods reduce the CPU time requirement by one to two orders of magnitude. However, the applicability of these algorithms to randomly heterogeneous soils with complicated geometry has not been studied.

Numerical Experiments

A number of numerical experiments were conducted using finite element techniques on the pressure(h)–based (5) and the mixed (2) form Richards' equation to examine the effect of time stepping, spatial discretization, mass lumping, initial, and boundary conditions on the accuracy, stability, and the convergence rates of these schemes. Whenever the simulation was limited to unsaturated conditions, the storativity was set to zero. Essentially, equations (3) and (1) were solved. However, for simulations involving layers and nearly saturated columns, a storativity of 1.0E–7 cm^{-1} was used. All numerical experiments were conducted using data reported by Zarba (1988). The soil column was 60 cm long and it was discretized to 30 line elements each 2 cm long. Linear basis functions were used for discretization. The saturated water content (θ_s) and the saturated hydraulic conductivity (K_s) were reported to be 0.3681 and 0.00922 cm/s, respectively. The constitutive relationship between pressure–water content and conductivity–saturation were expressed using the well known Mualem–van Genuchten (van Genuchten, 1980) relationship:

$$\theta_{(h)} = \theta_r + \frac{\theta_s - \theta_r}{(1 + |\alpha h|^{\beta})^{\gamma}} \tag{12}$$

$$K(S_e) = K_{S_e^{1/2}}\left[1 - \left(1 - S_e^{1/\gamma}\right)^{\gamma}\right]^2 \tag{13}$$

in which S_e is the relative saturation and θ_r is the residual saturation. The exponents γ and β are constants for a given soil type and the parameter α is inversely related to the bubbling pressure. S_e is defined as:

$$S_e = \frac{\theta - \theta_r}{\theta_s - \theta_r} \tag{14}$$

and γ and β are related according to $\gamma = 1 - 1/\beta$. For the given soil column the following values were used:

θ_r = 0.102
β = 1.9808
α = 0.03353 cm^{-1}

Effect of time step size on mass balance: For the pressure based (3) and mixed form (1) Richards' equation, the effect of time step on mass balance was calculated. The globally mass–lumped version of (1) was used. For the test problem, an initial condition of –100 cm of uniform suction was considered for the soil column. The top boundary was a flux type (0.008 cm/s) and the bottom was Dirichlet (–100 cm). The convergence criteria in the Picard scheme were: residual pressure = 1.0E–6 cm and the relative pressure change between the iteration = 0.001 cm. At the end of each iteration (within a given time step), the maximum pressure change was checked against the relative convergence criterion. Similarly, the right hand side was checked against the residual pressure. When both were less than the set tolerances, the time step was marched forward. Time of simulation was limited to 40 time steps. The results of the simulation are presented in Table 1. As seen in this table, the mass balance ratios for all time steps (for which convergence occurred) in the mixed formulation was 1.0, whereas, the MBR varied from 0.758 to 0.99 for the h–based scheme between time steps of 5 and 0.5 s.

Effect of initial conditions: For a time step size of 0.5 s, the impact of initial condition on mass balance and convergence is presented in Table 2. A flux of 0.008 cm/s was applied at the top of the soil column and the bottom boundary was set equal to the initial condition. The storativity was set to zero. The pressure–based form showed poor mass balance (as low as 74%). Moreover, at higher tension convergence could not be attained in either case. Simulations at an initial moisture content of −100 cm were carried out for both schemes with time steps of 1.0 s. The MBR for the pressure–based form ranged from 0.918 to 0.99 with a cummulative value of 0.976. No convergence occurred for this size time step when the initial pressure was reduced to −1000 cm. However, for the same case with Modified Picard formulation, convergence occurred between 9 and 7 iterations and the MBR was unity.

Effect of mass lumping: Using the mixed–form Richards' equation (1), the effect of mass lumping on the rate of convergence was analyzed. The time derivative terms in the global matrix were diagonalized in the mass lumped formulation. Two types of mass–lumping procedures are normally followed: element–level lumping and global lumping. The following explains the differences between the two types of mass lumping. For an interior node, the equation contains three unknowns: the unknown value at the node of interest and at the two connecting nodes. In element lumping, the off–diagonal time derivative terms are simply moved to the center diagonal. However, they still contain coefficients that need to be evaluated at the connecting nodes. On the contrary, in global lumping, the off–diagonal terms are moved to the center diagonal and again digonalized for that node only. The number of iterations to satisfy the preset convergence criteria were used as the measure of the rate of convergence. The time step chosen for this simulation was 5.0 s and a total of 40 time steps were simulated. The initial condition was −100 cm uniform for the entire column length. The top boundary was flux type (q = 0.008 cm/s) and the bottom was Dirichlet (−100 cm). The results are presented in Figure 1. Similarly, the effect of the type of mass lumping on convergence rates is presented in Figure 2. In this figure, the time derivative terms in the global and the element matrices were digonalized. It was observed that the globally lumped matrix required a fewer number of iterations than the matrix in which mass lumping was achieved on element levels. The pressure profiles calculated by the two methods were almost identical.

Effect of bottom boundary on a nearly saturated column: The effect of the type of bottom boundary on the pressure profile of a nearly saturated column was tested (Figure 3). The initial condition was −1.0 cm of suction uniform in the column. The time steps were 10 s and the total time of simulation was 1000 s. A flux of 0.004 cm/s was applied at the top of the soil column. In one case, the bottom was assumed to be "free–draining" or "unit–gradient" (which is equivalent to second type (McCord, 1991)) and in the other case, it was kept at saturation (h=0 cm). The modified Picard method with global mass lumping was used in both the cases. The number of iterations for the two types of pressure profiles are shown in Figure 4. Both the methods required similar number of iterations per time step to converge. The moisture profiles in the two cases are quite different and this difference is primarily due to the boundary conditions. The problem for the second case (Dirichlet bottom boundary) was tested using the standard pressure (h) based formulation with global mass lumping. A MBR as low as 0.675 was observed for the first time step and the number of iterations for this time step was four times higher than in the mass conservative modified Picard formulation. However, the MBR for the pressure–based formulation improved to over 95 percent after 10 time steps.

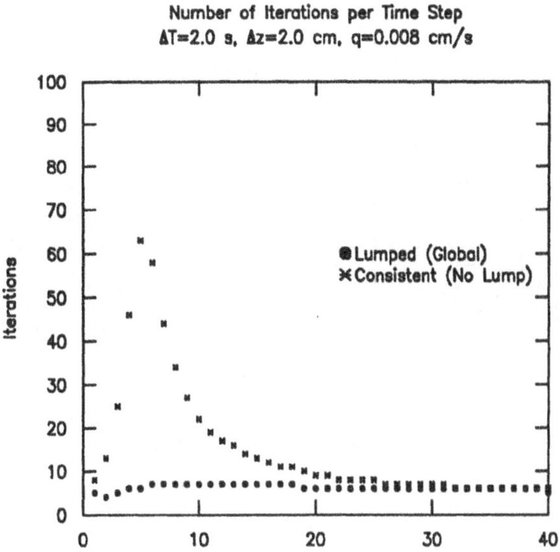

Figure 1. Effect of mass lumping on the rate of convergence

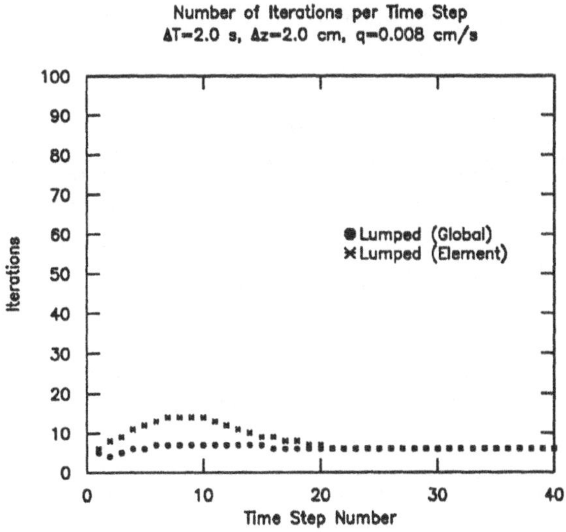

Figure 2. Effect of type of mass lumping on the rate of convergence

	h–based		mixed	
Δ t (s)	Range	Cumulative	Range	Cumulative
0.5	0.955–0.991	0.981	1.0 – 1.0	1.0
1.0	0.918–0.991	0.976	1.0 – 1.0	1.0
2.0	0.860–0.986	0.968	1.0 – 1.0	1.0
5.0	0.758–0.978	0.956	1.0 – 1.0	1.0
10.0	N.C.	–	N.C.	–

Table 1. Effect of time step size on the mass balance ratio between pressure–based (h–based) and the mixed form of Richards' equation.

N.C. – No Convergence

	h–based		mixed	
I.C. (cm)	Range	Cumulative	Range	Cumulative
–100	0.955–0.991	0.981	1.0 – 1.0	1.0
–1000	0.739–0.971	0.933	1.0 – 1.0	1.0
–2000	N.C.	–	N.C.	–
–5000	N.C.	–	N.C.	–

Table 2. Effect of initial conditions* on the mass balance ratio between pressure–based (h–based) and the mixed form of Richards' equation.

* based on a Δt of 0.5 s. No convergence for large time steps.

I.C. – Initial Conditions

N.C. – No Convergence

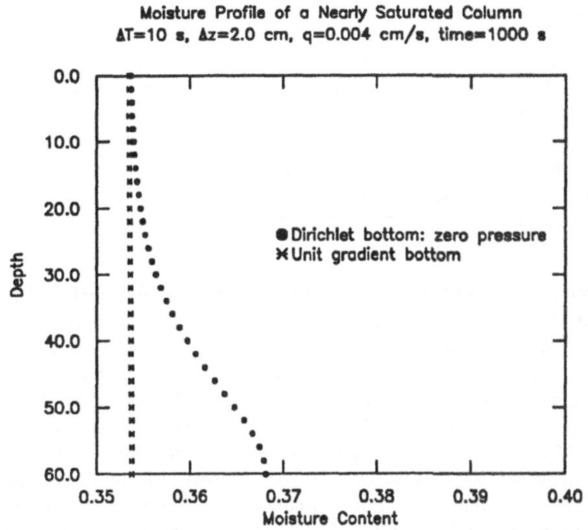

Figure 3. Effect of bottom boundary on pressure profile of an initially saturated column.

Note: Column bottom was set to zero pressure for Dirichlet condition.

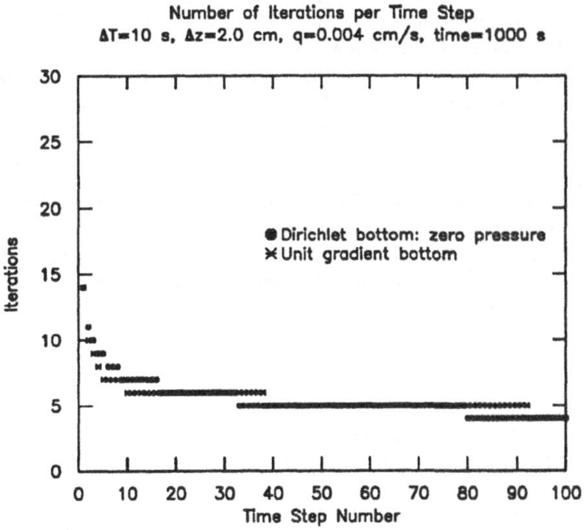

Figure 4. Effect of bottom boundary on the rate of convergence.

Effect of top boundary on moisture profile and convergence: The effect of the type of top boundary on pressure profiles is presented in Figure 5. In one case, a flux equivalent to the saturated conductivity (0.00922 cm/s) was applied and in the other case a Dirichlet condition (h=0 cm) was imposed. The initial condition was −100 cm uniform in the column and the bottom boundary was −100 cm (prescribed head). Simulation was carried out for 40 time steps using a time step of 5 s. The moisture front penetrated deep with the Dirichlet boundary condition compared to the saturated flux condition. The flux boundary required a larger number of iterations in the beginning. With the progress in time, however, this number was lower than the Dirichlet condition (Figure 6). This suggests that a flux equivalent to saturated conductivity should not be considered as the same as a Dirichlet boundary with zero pressure at the top.

Effect of time varying top boundary conditions on type of formulation: The moisture profiles and the convergence rates between the pressure–based and the mixed form Richards' equation were compared in Figures 7 and 8. The initial condition was −100 cm of suction uniform for the soil column. Time steps of 5 s were used in simulation. The bottom was a Dirichlet (h = −100 cm) boundary. At the top boundary, from the start at t=0 to a time of 50 s, a flux of 0.009 cm/s was applied; from 50 to 100 s, a Dirichlet condition (h=0 cm) was imposed, and finally from 100 to 200s, a no flux condition (q = 0 cm/s) was specified. Mass lumping (at global matrix) was applied to both schemes. The pressure profiles in Figure 7 are nearly identical in the two methods, however, the number of iterations (Figure 8) for the pressure–based formulation was substantially higher than the mixed form for the first three time steps. Especially, a jump in the number of iterations was observed for the pressure–based form when the top boundary type was changed from specified head (h=0 cm) to no flux type (at 100 s).

Effect of Time Step Control on Total CPU Usage: The effect of adaptive time stepping on CPU usage for long simulations was evaluated. The test problem was used. A flux of 0.0001 cm/s was applied at the top of the colum ad the bottom was set a fixed pressure of −100 cm. Total simulation time was approximately 31,000 s. Fixed time steps of 200 s were used in the first case. A total of 157 time steps were needed and the CPU time requirement was 53 s on a Hewlett Packard series 425 workstation. The number of iterations ranged from 8 to 3. In the second case, the time step size was allowed to drop by 50 percent if the number of iterations exceeded 20. Similarly, the time step size was allowed to increase by 20 percent if the number of iterations fell below 5. The total CPU use for the second case was 36.5 s, allowing for nearly 30 percent saving in computer time.

Effect of Soil Layering on Solution: The impact of a layered soil profile on the water content and convergence rates for the mixed form and the pressure–based equation was evaluated. A sandy–clay layer of 12 cm, at a depth of 24 cm from the surface was simulated. The saturated hydraulic conductivity of the clay was 0.0003693 cm/s and the saturated water content was 0.391. The van Genuchten hydraulic parameters were the following:

$\theta_r = 0.1$
$\beta = 1.48$
$\alpha = 0.059 \ \text{cm}^{-1}$

A flux boundary condition of 0.005 cm/s was imposed at the top. The bottom was fixed pressure of −100 cm. Simulation was carried out for 100 time steps each of size 10 s. The moisture front reached the layer interface and began saturating the top sand. The water content profiles for the two schemes are similar (Figure 9). However, the pressure–based scheme required significantly higher number of iterations at the begining (Figure 10). The MBR for the pressure–based scheme ranged from 0.744 to 1.018 with a cummulative value of 0.937. The mixed formulation had perfect mass balance.

Steady–State Simulation: The code was developed for transient simulationw where changes in pressure heads between the iterations (hence time steps)were the unknowns in the solution of the

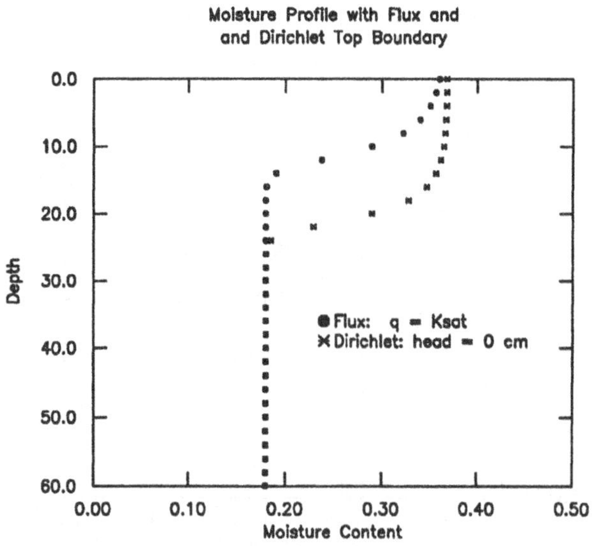

Figure 5. Effect of top boundary type on moisture content.

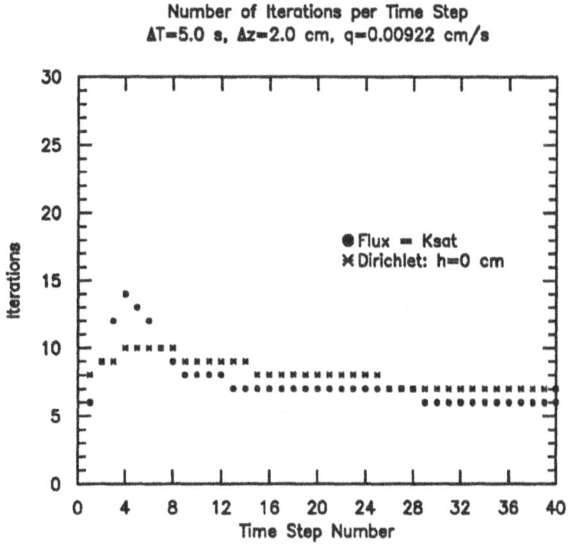

Figure 6. Effect of top boundary type on convergence rate.

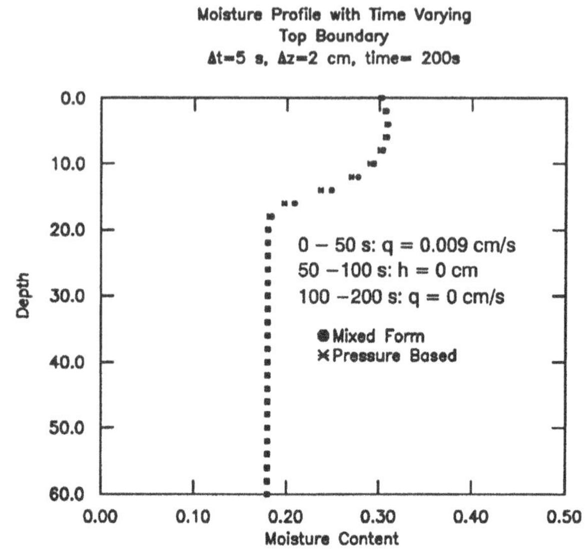

Figure 7. Effect of time varying top boundary on moisture content between the two schemes.

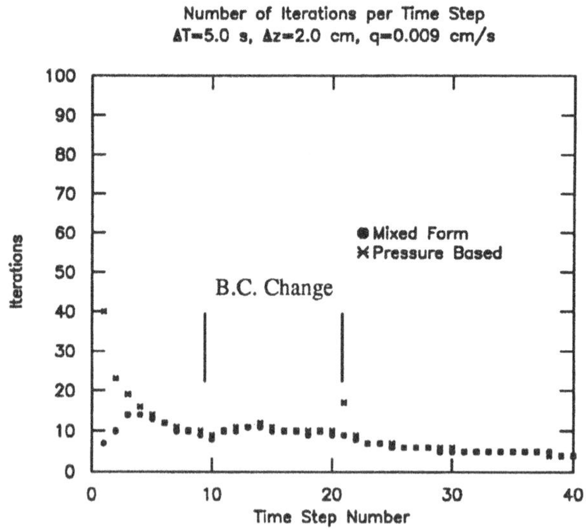

Figure 8: Effect of formulation on the rate of convergence for time varying top boundary.

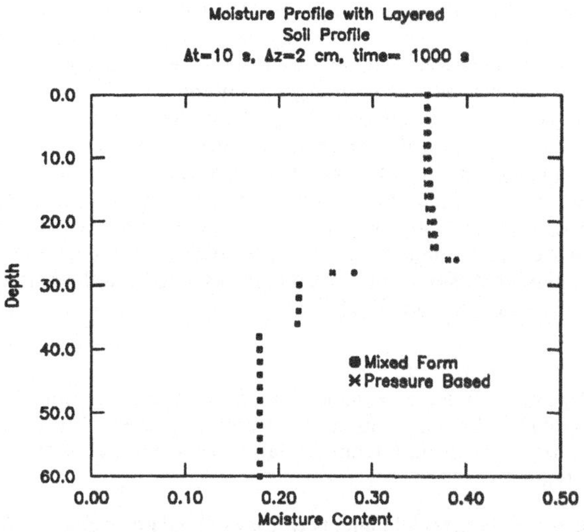

Figure 9. Effect of soil layering on moisture content between the two schemes.

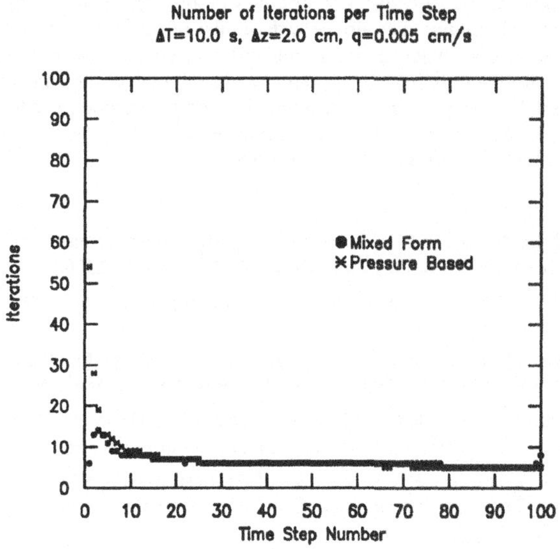

Figure 10: Effect of formulation on the rate of convergence for layered soil profile.

system of equations. In the Picard iterative solution technique used in this code, the new values of the pressure heads at the start of a time step are assumed to be equivalent to the old values and the nonlinear coefficients (K, C, and θ) are computed based upon these pressure heads. The iterative solution continues until the residuals meet the convergence criteria at each node point and the time step advances to the next. However, this solution technique produces a indeterminate system of equations for steady–state simulation since the difference between the old and new pressure head values are zeros. The ideal way to simulate steady–state flow is to remove the time derivative term from the flow equation and recode the problem. Since this requires the development of a new computer code, an easy way to get around this problem is to always guess the new pressure heads some thing other than the old pressure heads at the start of simulation. This will introduce some round–off errors, but prevent the formation of an inderminate system of equations. It was found that the round–off errors were insignificant to affect the accuracy of the result.

Conclusions

The mixed form of Richards' equation is preferred over the pressure–based form for simulating variably saturated flow because of its mass conservative property. The time derivative terms in the finite element matrix should be mass lumped (diagonalized) at the global level in order to reduce the total number of iterations.

The selection of the proper boundary type is essential for accuracy. The top can be any type of boundary, however, careful analysis is needed in selecting the bottom boundary. If the initial condition is hydrostatic (e.g. pressure decrease corresponds directly to depth) with the bottom being saturated, a Dirichlet bottom boundary with zero pressure may be the logical choice for infiltration problem. A unit gradient bottom boundary may also provide similar results. If the pressure at the bottom is set to other than 0 cm, large mass balance errors could be expected once the front reaches the bottom of the column Similarly, for the draining of an initially saturated column, the unit gradient bottom boundary is ideal. It is possible to run the transient simulation computer code (which uses the Picard iterative solution technique) for steady–state simulation by guessing the new pressure head at each node as some thing other than the old pressure–head values.

References

Celia, M.A., E.T. Bouloutas, and R. Zarba, 1990, A general mass–conservative numerical solution for the unsaturated flow equation, *Water Resour. Res.*, 26: 1483–1496.

Childs, E.C. and N. Collis–George, 1950, The permeability of porous materials, *Proc. Royal Soc. London Ser A.* (201): 392–405.

Davis, L.A. and S.P. Neuman, 1983, Documentation and user's guide: UNSAT2 – variable saturated flow model, NUREG/CR–3390, U.S. Nuclear Regulatory Comm., Washington, D.C.

Hills, R.G., I. Porro, D.B. Hudson, and P.J. Wierenga, 1989, Modeling one–dimensional infiltration into very dry soils: 1. Model development and evaluation, *Water Resour. Res.*, 25: 1259–1269.

Huyakorn, P.S., S.D. Thomas, J.W. Mercer, and B.H. Lester, 1983, *SATURN: A Finite Element Model for Simulating Saturated–Unsaturated Flow and Radionuclide Transport*, Electric Power Research Institute, Palo Alto, CA.

Huyakorn, P.S. J.B. Kool and J.B. Robertson, 1989, *VAM2D – Variably Saturated Analysis Model in Two Dimensions* (version 5,0 with hysteresis and chain decay transport): Documentation and User Guide, NUREG/CR–5352, HGL89–01, Nuclear Reg. Comm., Washington, D.C.

Kirkland, M.R., R.G. Hills, and P.J. Wierenga, 1992, Algorithms for solving Richards' equation for variably saturated soils, *Water Resour. Res.*, 28: 2049–2058.

Kool, J.B. and M.Th. van Genuchten, 1991, HYDRUS: *One–Dimensional Variably Saturated Flow and Transport Model, Including Hysteresis and Root Water Uptake*, U.S. Salinity Laboratory, Riverside, CA.

McCord, J.T., 1991, Application of second–type boundaries in unsaturated flow modeling, *Water Resour. Res.*, 27: 3257–3260.

Milly, P.C.D., 1985, A mass–conservative procedure for time stepping in models of unsaturated flow, *Adv. Water Resour.*, 8: 32–36.

van Genuchten, M. Th., 1980, A closed–form equation for predicting the hydraulic conductivity of unsaturated soils, *Soil Sci. Soc. Am. J.*, 44: 892–898.

Zarba, R.L., 1988, *A Numerical Investigation of Unsaturated Flow*, M.S. thesis, Dept. of Civil Eng., Mass. Inst. of Technology, Cambridge, Mass.

IS IT CORRECT TO ASSUME THAT MOVEMENT OF WATER IN UNSATURATED SOIL STRATA IS A PISTON FLOW

B.P. SINGH
Nuclear Science Centre, Aruna Asaf Ali Marg, New Delhi

BHISHM KUMAR
National Institute of Hydrology, Roorkee (U.P.), India

SYNOPSIS

Movement (flow) of water (moisture) in unsaturated soil strata depends upon the replacement of "air-water" held in unsaturated zone. If the soil strata is homogeneous from surface to water table, it is simple to work with, but in nature this may not be possible. The soil strata may be built-up of different layers of different compositions, compaction and of different grain size which may complicate the situation. Recharge to groundwater from surface is assumed to be a piston flow. This piston flow movement of water had been used to find the recharge using tritium tagging method. In order to study the piston flow, two experimental studies were conducted (1) using Ra-Be neutron probe and (2) gamma-ray transmission method using two bore holes under ponding conditions at the same site. The results by two methods are practically same for the water held and water drained with time. The authors (1,2) considered and proposed "column approach" for rate of discharge of moisture above field capacity by introducing two parameters, the residence time (τ) and movement of water front

V. P. Singh and B. Kumar (eds.), Subsurface-Water Hydrology, 65–76.

(T).

The data in these two experiments were re-examined to understand the piston flow model used in tritium tagging method. In this analysis, the movement is like stretching of water pulse applied at the surface throughout the length of the column. This will depend on the residence time (τ) which is governed by capillary forces, due to which the moisture is retained at field capacity. T and τ had been worked out for different ponded heights and at different depths of soil strata. For smaller ponded height, T decreases linearly (below 2 m) with depth. The flux of moving front at the slow rate also decrease with time giving the movement layer by layer. The quantitative estimates are made and discussed on the basis of this model.

INTRODUCTION

Movement of water in unsaturated soil strata depends upon the replacement of air-water in unsaturated zone. If the soil strata is homogenous from surface to water table, it is simple to work with, but in nature this may not be possible. The soil strata may be built-up of different layers of different compositions, compaction and different grain size which may complicate the situation. The piston flow movement of water had been used to find the recharge using tritium tagging method. Is this approach a model or the movement is actually piston flow ? What is the role of water held at field capacity in the process of movement of water to recharge groundwater ? Further, movement of irrigated water in the field (ponded water) is needed to be reinvestigated so as to understand the recharge to groundwater. This will also reveal movement of water (pulse) due to rainfall in unsaturated soil strata.

The movement of water (pulse) in unsaturated soil strata can be studied using thermalization of neutrons (using neutron moisture depth probe) and gamma ray transmission (or attention) methods as given by authors (1,4,5 & 7).

NEUTRON MOISTURE PROBE

Neutron moisture probes are readily available and can be used after the calibration in situ conditions. The determination of moisture is useful, but its real use can be if the change in moisture at a given site (in situ) in a bore hole is situated at different depths with time and followed for a full monsoon season. In order to find the applicability of this method, Roorkee group (1) had conducted an experiment which consisted of recording of neutron counting rate (i.e. moisture determination in a bore hole with aluminium casing in the initial soil conditions at various depths and then observing neutron counting rate under ponding conditions with respect to time until initial moisture conditions are attained). Neutron counting rate is transformed in the moisture content and analysis is done to understand the variation of soil moisture in a column (approach) and to understand the contribution of ponded water to water table. The initial moisture content held up due to field capacity plays an important role in the movement of moisture content. A typical study reported by Roorkee group (1), supplied water at the rate of 81 c.c/sec and filled the pond upto to 21.67 cm in 65 minutes. Many sets of observations of neutron content rates were taken at the interval of 30 minutes for few and later on at the intervals of 1-5 hours. The soil moisture variations with depth and time was recorded. Using these observations, soil moisture, was integrated in a column up to

3.0 m, 2.5 m, 2.0 m, 1.5 m and 1.0 m as given in figure 1. These graphs did not reveal any pattern of movement of soil moisture. As the moisture held up at the field capacity at various depths (known as initial moisture content, plays an important role, therefore, this initial moisture was subtracted and then the moisture content was integrated in the column up to 3 metre. Figure 2 gives the variation of such integrated moisture content in column with depth and time. Smooth variation and exponential form of curves of integrated moisture content are quite interesting to develop mathematical model of the soil water system.

GAMMA-RAY TRANSMISSION (ATTENUATED) METHOD

Gamma-ray scattering probes are developed to find the density. This technique also did not find much use in situ. This consists of strong source of gamma-rays and detector in the same sonde. If we separate the source and detector and place few centimetres apart and use it for absorption/attenuation of gamma rays due to soil, it will provide the information of density and also change in density due to moisture in soil water system. If we make two bore holes, one for lowering of source and other for lowering the detector simultaneously, we can determine the change in soil moisture with depth and time.

Attenuation of gamma-rays through soil water column is worked out by Singh and Kumar (3). In the soil column approach, the change in logarithm of counting rate is directly proportional to the change in moisture content, and hence the change in moisture content is determined in the different sections of the soil column and is summed up to full length of the column, it will amount to total water applied at the surface of the soil column. Singh and

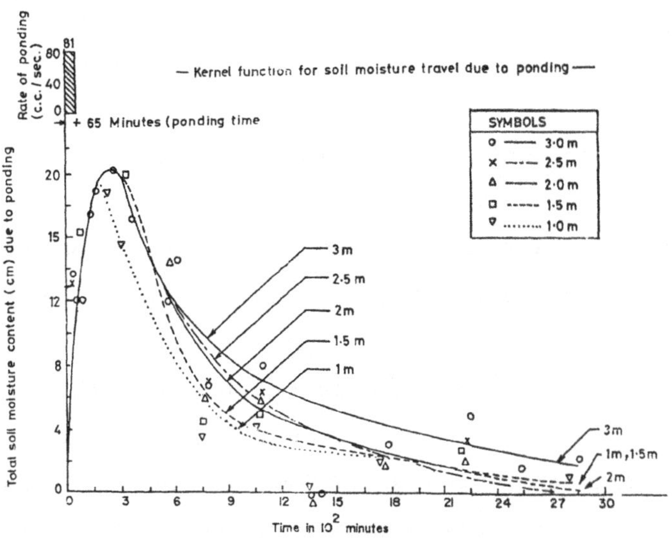

FIG.1 TOTAL SOIL MOISTURE CONTENT DERIVED FROM INTEGRATING SOIL MOISTURE PROFILES OVER DIFFERENT DEPTHS INDICATED. TOTAL PONDED WATER WAS 21.67 cm (SURFACE DEPTH) OVER THE AREA

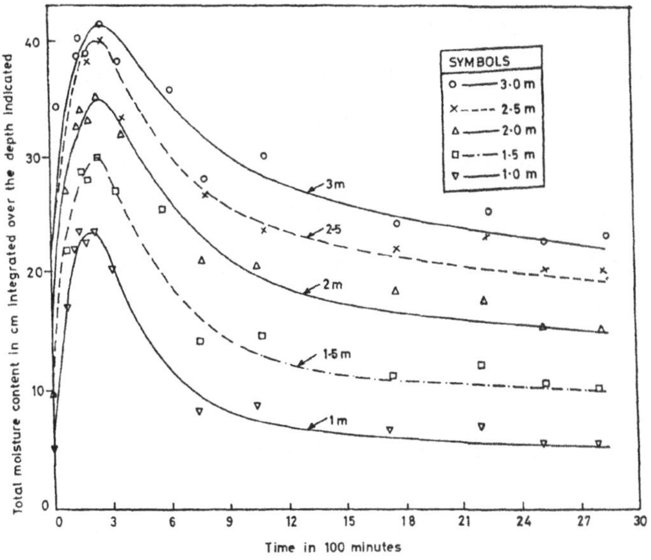

FIG.2 TOTAL SOIL MOISTURE CONTENT DERIVED FROM INTEGRATING SOIL MOISTURE PROFILES OVER THE DEPTHS INDICATED WITH FIELD CAPACITY SUBTRACTED

Chandra (5), and Kumar and Singh (3) have reported that the factor $\{(\mu_w / \rho_w) \rho_w c x\}$ is constant as given in the equation,

$$\sum_{0}^{L} (\ln N_{sw} - \ln N'_{sw}) = (\mu_w/\rho_w)\rho_w x c(\theta'_v - \theta_v) \quad (1)$$

where N_{sw} and N'_{sw} are counting rates corresponding to the initial moisture content (θ_v) at time $t = t_0$ and the moisture content (θ'_v) after a certain interval of time $t = t_1$ after applying the water at the surface respectively, on volume basis and μ_w, ρ_w are mass attenuation co-efficient and density of water, c is a constant, x is the thickness of soil column, and L is length (depth) of soil strata. The summation is for full length (L) of the soil column i.e., natural logarithm of the counting rate is multiplied by the length of soil column under consideration and summed up over such readings. Therefore, the summation of $(\theta'_v - \theta_v)$ is the total amount of water applied at the surface of the soil column at time $t = t_0$ and observed at $t = t_1$. The factor, $\{(\mu_w / \rho_w) \rho_w c x\}$ can be taken as calibration factor, provided x is constant, and is determined as suggested by Kumar and Singh (2).

The calibration approach of gamma ray transmission system in field is slightly different from laboratory approach as reported by Kumar and Singh (2). In the field, two bore holes cannot be made exactly parallel. Hence, the variation of the distance between source and detector was also taken into account in the calibration of gamma-ray transmission system in situ.

In a typical study carried out in the field, the distance between source and detector varied at the rate of 0.20 cm per 2.5 cm depth. Thus it was 37.7 cm at the ground surface and 35.5 cm at the depth of 3 m below ground surface. In the present situation, the equation (1) can be written as,

$$\sum_{0}^{L}(\ln N_{sw}-\ln N'_{sw})/X=(\mu_w/\rho_w)c\sum_{0}^{L}(\theta'_v-\theta_v) \quad (2)$$

i.e., the difference of the logarithm of counting rates is divided by the thickness of soil medium at each stage of observation and is summed up instead of summing of difference of logarithm of the counting rates. In the present study, the calibration of gamma-ray transmission system was done in field by supplying known volume of water, i.e. height of water at the surface of area under study. Initial gamma-ray counting rates were note down at various depths. After complete absorption of supplied water into the soil media, gamma ray counting rates were again noted down at various depths and at different time intervals. The change of counting rates is summed up after taking logarithm and dividing by the distance between source and detector (thickness of soil medium) at that particular depth of observation. The total change of counting rates is compared to the total irrigated water (using equation 2) and the calibration constant was determined each time.

Change in water content between source and detector due to water applied and as observation by gamma-ray attenuation reveals an interesting approach, i.e., the rate of change of water content in the soil column (between source and detector) as shown in figures 3, 4 and 5, can be used to understand the movement of water flux V(z). Continuity equation relating to the change in water content to the water flux, V(z) is given by

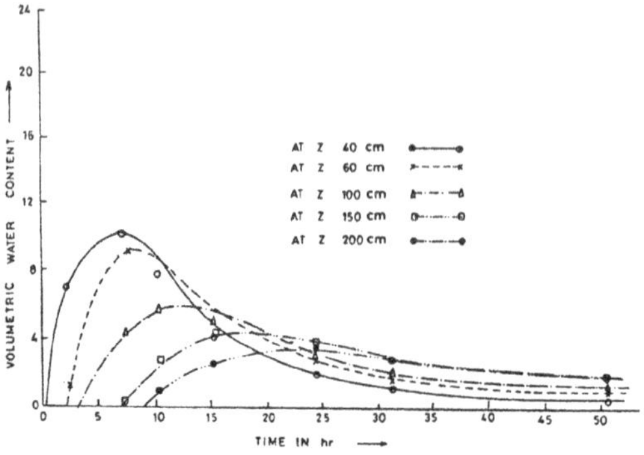

FIG.3. VARIATION OF VOLUMETRIC WATER CONTENT WITH TIME FOR 10 cm EQUIVALENT HEIGHT OF WATER APPLIED IN FIELD

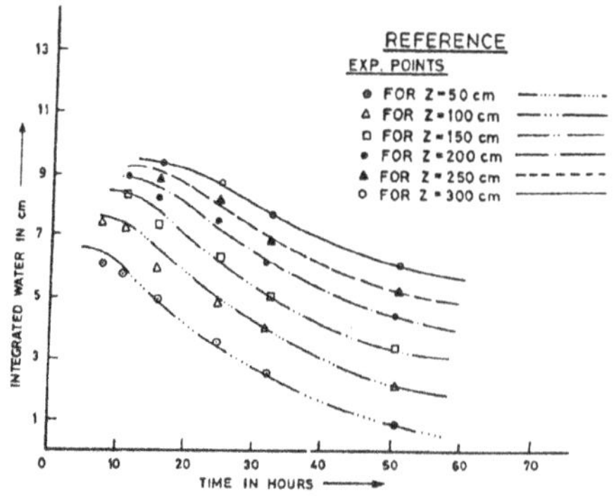

FIG 4 - VARIATION OF INTEGRATED WATER CONTENT WITH TIME FOR 10 cm EQUIVALENT HEIGHT OF WATER APPLIED IN FIELD

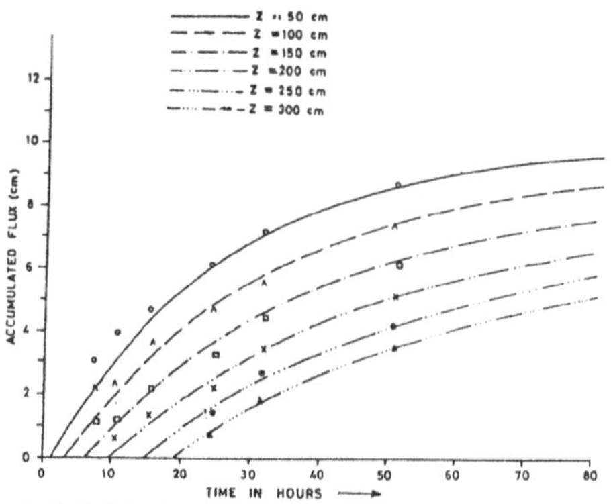

Fig. 5 VARIATION OF ACCUMULATED FLUX WITH TIME THROUGH VARIOUS LENGTHS OF THE SOIL COLUMN FOR 10 cm INSTANTANEOUS PONDING OF WATER IN FIELD

$$\frac{\partial \theta}{\partial t} = -(\partial V/\partial z) \qquad (3)$$

where θ is the volumetric moisture content and t is the time. On integrating between depths, Z = 0 to Z = L

$$V(L) = \int_0^L (\partial \theta/\partial t)\, dZ$$

(L) can be known at any time.

The accumulated flux F(Z) passing through (Z) length of the soil column between times t_1 and t_2 can be given mathematically by the following equation:

$$F(Z) = -\int_{t1}^{t2} V(Z)\, dt = \int_0^Z (\theta_{(t1)} - \theta_{(t2)})\, dZ \qquad (4)$$

$$F(Z) = \int_0^Z \theta_{(t1)}\, dZ - \int_0^Z \theta_{(t2)}\, dZ \qquad (5)$$

t is difficult to obtain moisture flux V(Z) by the equation (4) But V(Z) can be obtained with the use of equation.

$$\partial F/\partial t = -V(Z) \qquad (6)$$

The plot of integrated flux with time through different lengths of the column can be made.

In this method, the change in moisture content above the field retension capacity (initial soil moisture) has been considered.

Figure 3 gives the variation of volumetric moisture content in a depth at Z = 40 cm to Z = 200 cm with time for 10 cm equivalent height of water supplied in the experimental field in a typical studies carried out by Roorkee group (6). In this study two bore holes each of 10 cm. diameter upto depth 3 m, 37.9 cm apart, with aluminium casing were used. NaI (Tl) scintillator detector with spectrometer and 50 mci Cesium-134 radioactive source were housed in these two bore holes. These source and detector both can be lowered keeping the relative position fixed.

The variation of integrated moisture in a column of various depth is given in figure 4 and variation of accumulated flux with time, through various length of soil column, is given in figure 5. Further, the time differential of this curve F(Z) i.e. V(Z) are obtained as given in figure 6.

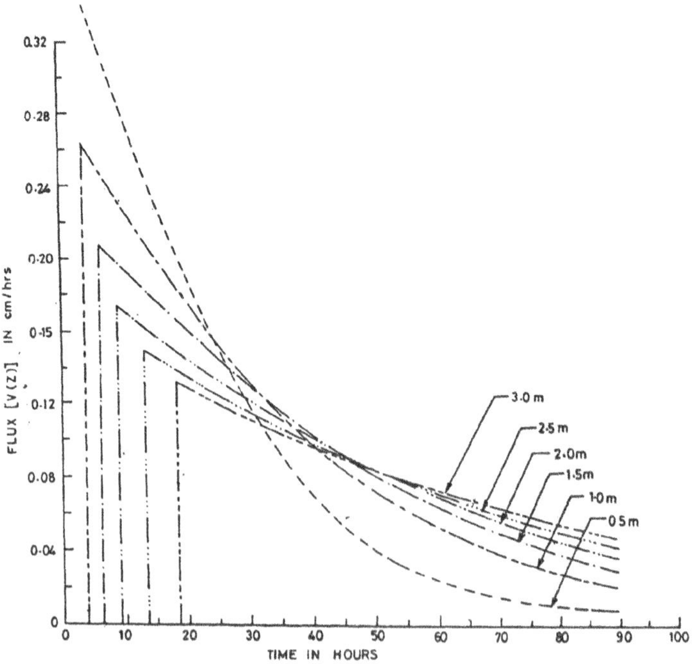

FIG. 6 VARIATION OF FLUX |V (Z)| WITH TIME THROUGH DIFFERENT LENGTH OF THE SOIL STRATA FOR 10 cm PONDING OF WATER IN FIELD

There are two components in these curves (1) an extrapolation as given in figure 5, gives the time arise which may be taken as movement of water front (say T) (ii) an extrapolation for infinite time which will give the amount of water supplied (ponded) i.e. the time differential portion of these curve as given in figure 6. This portion may be replotted, taking logarithm of the water (flux against time, giving a parameter, the reciprocal of which is in the unit of time, may be taken as mean residence time (τ). As such, the movement of water front (T) and mean residence time (τ) are two parameters which can be worked out for different pondings in the field.. These two can be used for calculation of movement of water in the unsaturated soil strata due to irrigation and rainfall.

The mean residence time for vcarious depths of ponded water is plotted with depth as given in figure 7. The slope of this curves at a depth beyond 1-1.5 m may be taken as straight lines.

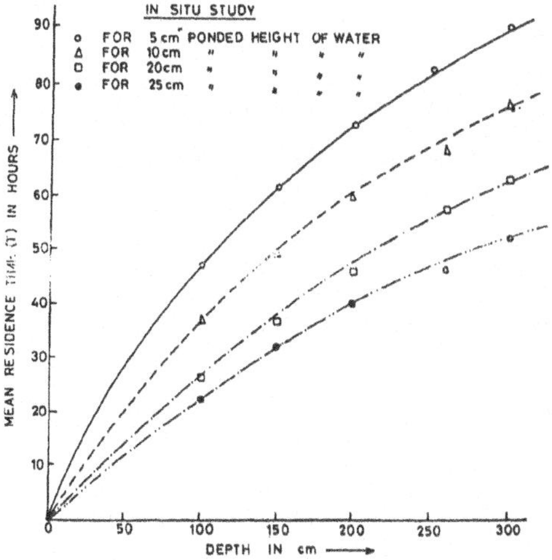

FIG.7. VARIATION OF MEAN RESIDENCE TIME (T) WITH LENGTH OF THE SOIL COLUMN FOR DIFFERENT PONDED HEIGHTS OF WATER

CONCLUSION

It is found that

(a) Water held at field capacity is important as concluded from figure 2 using neutron probe.

(b) Movement of water in unsaturated zone is like stretching the water pluse with time.

(c) The flow of water may be taken as, discharge from one "column" to another "column" and within the column, there may be dispersion.

ACKNOWLEDGEMENT

The authors are thankful to the authorities of Physics Department, University of Roorkee and National Institute of Hydrology, Roorkee, India for providing the facilities to carry out the experimental work.

REFERENCES

1. Kiran Shankar, Bhishm Kumar and B.P. Singh Indian National Science Academy. Vol.45 part A (1979) P.476.484.

2. B.Kumar and B.P. Singh. International Journal of Applied Radiation and Isotopes, Vol.30, p.493.

3. B.Kumar and B.P.Singh. Indian Journal of Pure and Applied Physics, Vol.19, 1981, p.49.

4. B.P. Singh and Bhishm Kumar, Indian Journal of Pure and Applied Physics, Vol.26, 1988, P. 376.

5. B.P.Singh and Satish Chandra, Journal of Hydrology, Vol.37, 1977, p. 189.

6. Bhishm Kumar, B.P. Singh and Satish Chandra Institute of Engineers (India), Vol.73, 173-179 (1993).

7. W.H. Godner, Soil Science Society of America, method of Soil Analysis Part I, Physics and Menerological Methods - Agronomy Monography and Edition 493-544.

A NEW TECHNIQUE FOR THE CALIBRATION OF NEUTRON-MOISTURE PROBE

BHISHM KUMAR, Rm. P. NACHIAPPAN & RAJAN VATSA
National Institute of Hydrology, Roorkee-247 667 (UP), India

SYNOPSIS

A new technique for the in situ calibration of neutron moisture probe has been developed, which is very simple and accurate. In this technique, known amount of water (equivalent to some ponded height of water) is applied at the soil surface. The infiltrated water is accounted for by integrating the change in observed volumetric moisture content by neutron probe. This total infiltrated water should be equal to the water applied at the surface owing to the fact that evaporation loss is negligible. A simple linear regression analysis of the applied water at the ground surface and observed moisture in the soil-system using neutron probe provides the calibration factor. The experiment carried out in the NIH campus for the calibration of neutron soil moisture probe is reported. The results reveal the better performance of the new technique over gravimetric calibration technique.

INTRODUCTION

Neutron moisture measurement technique is now an established one in hydrological investigation as it provides in situ and instantaneous information of soil moisture. Its major applications are in soil water storage and water budget computations. The probes, though very much convenient for repetitive moisture measurements may lead to erroneous interpretations, if they are

V. P. Singh and B. Kumar (eds.), Subsurface-Water Hydrology, 77–87.

not properly calibrated. Though, commercial probes are now supplied with factory calibrated curves, specific calibrations are required for different soil types. It is also evident that since the soil may consist a range of chemical elements some of which may be strong absorbers (e.g. Cd, Cl, Fe, etc.) each soil has a unique calibration curve. The techniques currently employed for the calibration of the probes viz. theoretical calibration using soil chemistry, laboratory drum calibration or field gravimetric calibration do not provide the accurate calibration. This is due to inaccurate determination of dry density at deeper depths in situ and also extensive and laborious sampling procedures including tedious chemical characterization of the soils are involved with these techniques. A new technique for the calibration of the neutron soil moisture probes has been attempted. A similar approach has been tried successfully and reported earlier for gamma ray transmission method (Bhishm Kumar and Singh, 1979).

THEORETICAL ASPECTS

A neutron moisture probe generally consists a fast neutron source and an adjacent slow neutron detector. The source (as Americium 241: Beryllium in Troxler Model 3320 series) emits fast neutrons into the surrounding soil. These neutrons collide with the nuclei of the soil atoms, predominantly with those of hydrogen in soil water, loose their kinetic energy and gets slowed down. Thus a cloud of slow neutrons is generated within the soil strata around the source. The density of this cloud is largely a function of soil water content and is directly proportional to the hydrogen nuclei concentration of the soil water molecules. The thermalised (slowed down) neutrons are detected by the detector (such as

Helium - 3 in case of Troxler Model 3320 series).

The gauge is calibrated to get a relationship between actual moisture content and moisture read by the probe. Once this relationship is established for particular soil type the probe allows repetitive soil moisture measurement without resorting to laborious techniques.

CALIBRATION OF NEUTRON PROBE

The calibration of the probe is essential to establish the exact relationship between the moisture count of the probe and the actual moisture content in the soil media under consideration. Several studies have been reported on the effect of soil density (Holmes, 1966; Cole and Green, 1969), soil texture (Lal, 1979) on the above mentioned relationship. Though several methods have been proposed by various researchers (Holmes, 1966; Vachaud et al., 1977) the gravimetric method seems to enjoy a prominent place among the users, in spite of its inherent disadvantages. This is probably because the alternative methods proposed involve more laborious and precise estimation of certain parameters.

Calibration Using Gravimetric Techniuque

The gravimetric technique can be used to calibrate the probe either in laboratory or in field. This method is supposed to be the simplest and easier method of calibration. The calibration of probe in field involves the collection of soil samples from different depths by making a bore hole of desired diameter and measurement of moisture content using infrared moisture balance or oven drying method. Temporary access tube is then installed in the same auger hole and neutron counts or directly the moisture contents are obtained using the probe. Linear regression is performed on the moisture content determined using gravimetric technique and the corresponding count ratios/moisture content obtained by using the neutron probe.

Problems Associated

1. Due to soil heterogeneity and various sampling errors there is often a wide scatter in the calibration points.

2. Soil samples are to be collected at least from six points to reduce the errors (Bell, 1976).

3. Since it is recommended that the analysis should be carried out under a range of soil moisture content, it usually takes a year or so to finalise the calibration (Ramesh Chand et al., 1986)

4. There is always certain possibility of error by the introduction of a common bias in the measured moisture content. All soil corers tend to compress the soil to some extent, unless enough care is taken, resulting in considerable errors in the estimation of volumetric moisture content.

5. The more prominent error occurs in the determination of dry bulk density. Special type samplers have to be used for the purpose which calls for digging pits for more accurate results, which again is laborious and time consuming. Further, even a ±0.1% error in dry density can result in serious errors in moisture determination (Cole & Green, 1969).

6. Moisture estimated using probe which was calibrated using gravimetric technique showed large systematic mean deviation of -25% in comparison to weighable lysimeters (Shude, 1983).

NEW TECHNIQUE FOR CALIBRATION

The calibration of the moisture probes essentially involves accounting for the soil moisture present in the soil column through the probe counts. This is possible by carrying out a linear regression analysis on the probe data and the actual soil moisture. It is very difficult to determine the latter without disturbing the soil column. Further it leads to errors in the determination of dry bulk density as pointed out in the preceding section. A new technique has been successfully tried to calibrate neutron moisture probe as per details described below.

In this technique a known amount of water, equivalent to certain ponding height of water, is supplied at the ground surface having a boundary in order to check the surface runoff. The probe readings taken at different time intervals are considered as volumetric moisture contents calculated with slope 1 and intercept 0. The initial moisture content prior to the addition of water is deducted from the values obtained for any particular time interval. This difference in moisture content is plotted against the corresponding depth of the soil. The area under the curves for different set of readings at different times, is a function of the actual moisture present in the soil column. The curve area, which gives the total volumetric water content of the soil column at that particular time as measured by the probe, is then evaluated This plot will give the integrated moisture content of the soil column since the start of the exercise.

This exercise can be carried out for different height of ponding such as 4 cm, 5 cm, etc. The calibration factors m and c can be obtained through the linear regression of the integrated water content and the applied ponding height of water.

SOME CLARIFICATIONS

The question of accounting for the lateral movement of

moisture as a result of lateral flow has been raised by a few fellow researchers. Therefore two additional access tubes installed close to the boundary of the experimental area (Tubes B & C as shown in fig.1) were utilised to observe the change in moisture content due to lateral flow, using neutron probe. The readings were taken for the same time interval as for the main tube (A). It was found that there was negligible or no change in the readings of initial moisture content indicating that there was no apparent loss of applied water from the boundary. However, this can be understood in' a better way by the fact that the experimental area was of the size 2m X 2m and the bore hole was made at the centre of it. The G.I plates were used to make a 60 cm deep boundary around the experimental area. Therefore the supplied water will not have any direct access for the lateral flow. Moreover, the loss of supplied water is only possible from the boundaries after a vertical flow upto the depth of 60 cm. depth, thereby, the supplied water will be completely absorbed by the time it will reach upto this depth. Therefore either there will be no loss of the supplied water from the area near central point or it will be negligible.

ADVANTAGES

1. The calibration procedure is simple and involves less labour.
2. The technique can be applied for insitu calibration and hence is representative of the field conditions.
3. As the method involves relative measurements of the applied water in terms of moisture contents by the probe at different times, the effect of variation in soil chemistry and texture could be considered as nullified for the purpose of soil moisture

estimations.

EXPERIMENTAL VERIFICATION

The proposed new technique was applied to calibrate the neutron moisture probe Troxler model 3320 series on the sandy loam soil in the Institute campus at Roorkee. A 2m x 2m x 1m mild steel square type infiltrometer was installed, such that 60 cm of the height of the infiltrometer was below the ground level. An aluminum access tube (A) was installed at the center of the infiltrometer. Two additional access tubes (B & C) were installed at one meter distance from the infiltrometer as shown in figure 1.

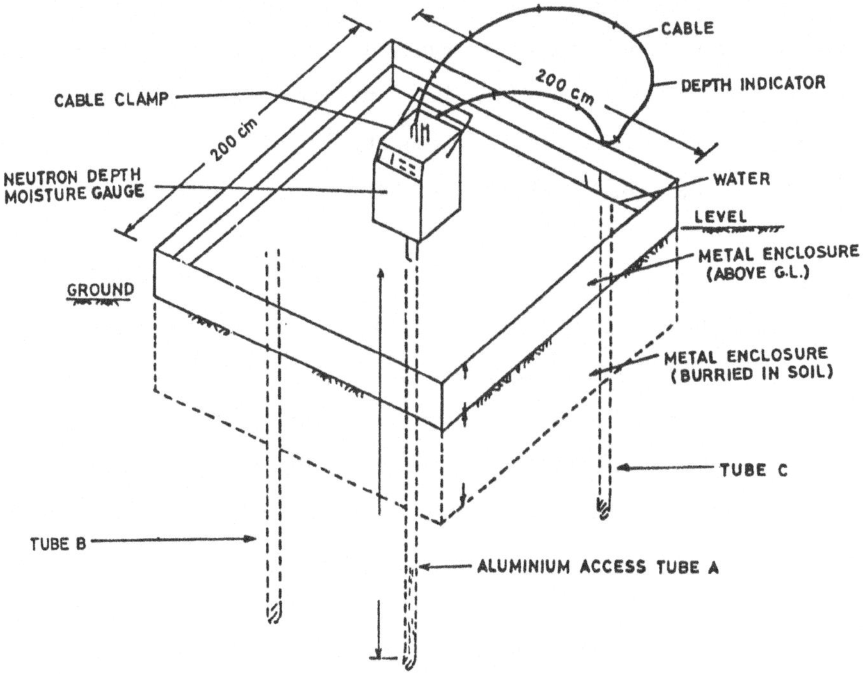

FIG 1—FIELD EXPERIMENTAL SETUP FOR DETERMINATION OF SOIL MOISTURE USING NEUTRON PROBE

Just prior to commencement of the actual exercise, probe readings were taken to calculate the initial moisture present in the soil column. The measurements were repeated at different time intervals to ascertain that the initial moisture content is at field retention capacity. An amount of water equivalent to a ponded height of 4 cm. was added to the infiltrometer. It was ensured that no additional water was introduced into the system other than this artificial ponding. There was no rainfall recorded at the site during the periods of these experiments. The probe readings were taken at different time intervals after the ponding. The recommended procedures were adopted for obtaining the counts. That is 30 sec. count times at every 10 cm interval. The count ratios thus obtained for each time interval was converted into volumetric moisture contents using the equation

$$\theta = m\ (A_0 + A_1\ X + A_2\ X^2 + A_3\ X^3) + c$$

where,

θ = volumetric moisture content

$A_0 - A_3$ = calibration coefficients (Factory calibrated)

X = count ratio

m = slope (= 1.0 for the purpose of calibration)

c = offset (= 0.0 for the purpose of calibration)

The initial moisture present in the soil system was deducted from the values thus obtained. The volumetric moisture contents were plotted against the corresponding depths, for each time interval (Fig. 2). The area under the curve was evaluated and it represents the total moisture content present in the soil system under study in units of depth. It is noticeable that the area under each curve are almost identical.

The same exercise was repeated for 5 cm ponding height of water (fig. 3). The calculated integrated moisture content were plotted against the corresponding height of water applied at the ground surface (Fig.4). The linear regression analysis on these data was performed. The calibration factors slope and the offset are 1.775 and 1.27

The details of water supplied at the surface and the integrated moisture content determined by summing the change in moisture content observed at different depths (area of the curve plotted between change in moisture content and depth at different times), as referred above, are given in Table - 1.

Table-1. Total soil moisture observed at different times for different ponding height of water water

Ponding Height (cm)	Integrated Moisture (cm) at different times (hours)			
4	Time elapsed	2.55	4.40	6.20
	Total moisture	1.4474	1.5546	1.5984
5	Time elapsed	4.15	6.40	9.15
	Total moisture	2.1305	2.0784	2.0669
10	Time elapsed	46.45	49.20	52.00
	Total moisture	6.9962	7.2650	7.0567
15	Time elapsed	45.00	50.15	52.45
	Total moisture	9.2902	9.3180	9.2142

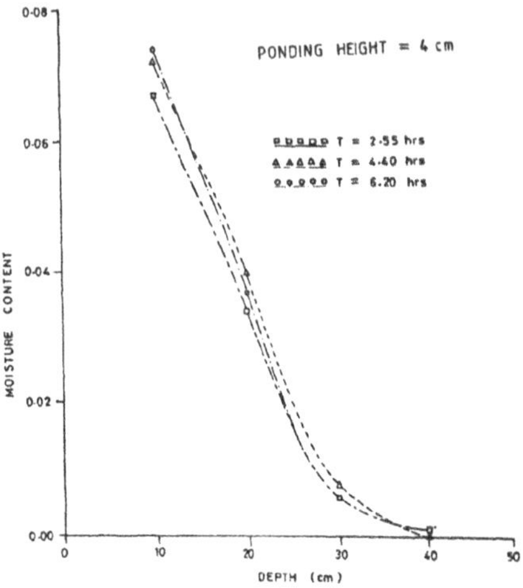

FIG.2 VARIATION OF MOISTURE CONTENT WITH DEPTH FOR 4 cm PONDING OF WATER

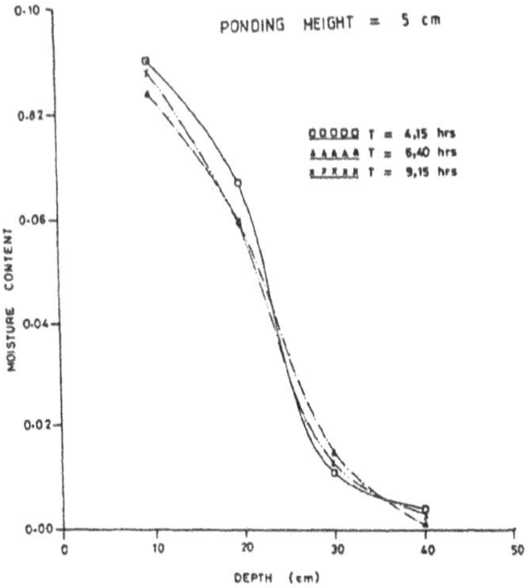

FIG.3 VARIATION OF MOISTURE CONTENT WITH DEPTH FOR 5 cm PONDING OF WATER

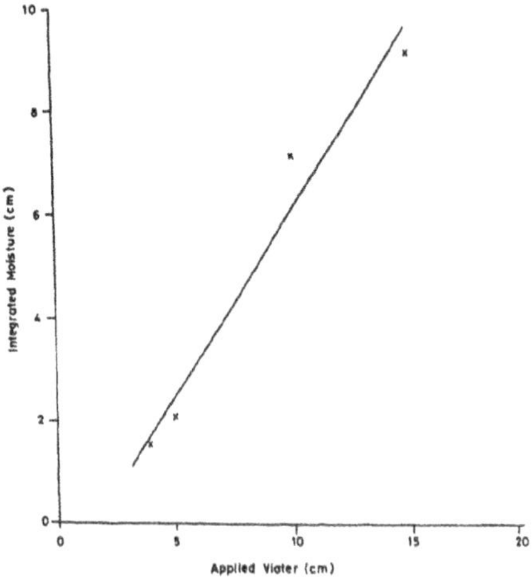

FIG. 4 VARIATION OF INTEGRATED MOISTURE WITH DIFFERENT PONDINGS OF WATER IN NATURAL FIELD

REFERENCES

Bell, J. P.(1976),'Neutron probe practice', N.E.R.C., Institute of Hydrology, Rep. No. 19, (IH, Wallingford, U.K.)

Bhishm Kumar and B. P. Singh (1979), 'A quick and accurate calibration of the - ray transmission method for the study of a Soil - water system', Intl. J. Applied Radiation and Isotopes, 30:493-495

Cole, J. A. and M. J. Green (1969), 'Measuring soil moisture in the Brenig catchment area: problem of using neutron scatter equipment in peaty areas', In: Water in the unsaturated zone (Ed. Rijtema, P. E. and H. Wassnik), Proc. Wageningen symposium, Unesco, Vol. I: 63-74

Holmes, J. W. (1966), 'Influence of the bulk density of the soil on neutron moisture meter calibration', J. Soil Sci., 26: 124-129

Lal, R. (1979), 'Concentrations and size of gravel in relation to neutron moisture and density probe calibration', Soil. Sci., 127:41-50

Ramesh Chand, S. M. Seth and Pratap Singh (1986),'Study of soil moisture using neutron probe', NIH, TN - 28

Schude, P.(1983),'The accuracy of measurements of soil water content made with a neutron - moisture meter, calibrated gravimetrically in the field', J. Hydrol., 62:355-361

Vachaud, G., J. M. Royer and J. D. Cooper (1977), 'Comparison of methods of calibration of a neutron probe by gravimetry or neutron-capture model, J. Hydrol., 34:343-356

SECTION 2:

SEEPAGE AND DRAINAGE

ESTIMATION OF CANAL SEEPAGE IN A LAYERED POROUS MEDIA USING BOUNDARY ELEMENT METHOD

by
T.I.ELDHO and B.VASUDEVA RAO
Department of Civil Engineering
Indian Institute of Technology
Powai, Bombay - 400 076.

ABSTRACT:

An estimate of the canal seepage losses from an unlined irrigation channel to the adjoining aquifer is important in the initial stages of canal design and also in the efficient utilization of available water. Here, a mathematical model based on the boundary element method is presented to estimate the steady state seepage losses from an irrigation canal into the adjacent layered isotropic media. The results are compared with those obtained by using the finite elment method and found to agree satisfactorily. The formulation of the problem using BEM is found to be simple and convenient compared to other methods of analysis.

INTRODUCTION:

In unlined irrigation canal systems, considerable amount of water is lost as seepage into the aquifer. The consequence of this loss of water is the rise in water table in the aquifer leading to water logging. A general study by several investigators indicated that this seepage loss some times may be as high as 50%. A specific case study relating to the Mahi Right Bank canal command area in Kheda district of Gujarat state (Rastogi and Joshi, 1991) revealed that nearly 0.2% of the total command area already became uncultivable due to water logging and 3.2% of the area is almost at the verge of water logging, and in about 30% of the area, water table is rising at alarming rate. Though the factors like poor drainage and return flow from excess irrigation are contributing partly to this phenomenon, the major cause of this problem is the heavy seepage from the canal. Therefore, a proper estimate of seepage from irrigation canals is important for the proper design, management and maintenance of the irrigation system.

Seepage losses from canals can be estimated either by direct measurement or by prediction. Direct evaluation of seepage from canals is based on obtaining discharge measurements at various

V. P. Singh and B. Kumar (eds.), Subsurface-Water Hydrology, 91–100.

points like inflow-outflow techniques, ponding method, tracer techniques etc. Estimation of canal seepage by prediction is based on the knowledge of relevant hydraulic properties and boundary conditions subject to hydrodynamic analysis.The different methods of seepage prediction include the closed form analytical solutions, flownets, electrical analogue models and numerical methods such as Finite Difference, Finite Element and Boundary Element methods. Numerical methods are being employed now-a-days on a large scale for solving complex ground water flow problems. Of the available numerical methods, Finite Difference and Finite Element methods are well established and are in common use. So here, an attempt has been made to apply the Boundary Element method to solve the canal seepage problem. In this method, the governing differential equation is transformed into a form of boundary integral equation. The main advantage of this method is that it reduces the computational dimensions of a problem by one, that is, a three-dimensional problem into a two-dimensional one and two-dimensional problem into a one-dimensional one, so the number of simultaneous equations to be solved is very much reduced. Other advantages are: ease in data handling, accuracy, efficiency, ability to solve the free surface problems with little grid generation, ease with which the infinite domain problems can be handled etc. [Brebbia, 1978].

FORMULATION OF THE PROBLEM:

Using the principle of continuity of flow and Darcy's law,the governing differential equation for steady flow through isotropic layered permeable medium, in two dimensions can be written as [Jacob Bear 1979]:

$$\frac{\partial}{\partial x}\left(K\,\frac{\partial \Phi}{\partial x}\right) + \frac{\partial}{\partial z}\left(K\,\frac{\partial \Phi}{\partial z}\right) = 0 \tag{1}$$

where ϕ is the potential function or hydraulic head, K is the permeability of the porous medium considered. As shown in Fig. 2, the porous media below the canal is a layered one, so that for each layer K is treated as constant and the Laplace Equation

$$\nabla^2 \Phi = 0 \tag{2}$$

is valid for each layer. If there are two layers, then the potential ϕ is constant at the intersection of two layers, that is

$$\Phi_{layer\ 1} = \Phi_{layer\ 2} \tag{3}$$

Also, at the intersection of layers, the amount of fluid leaving

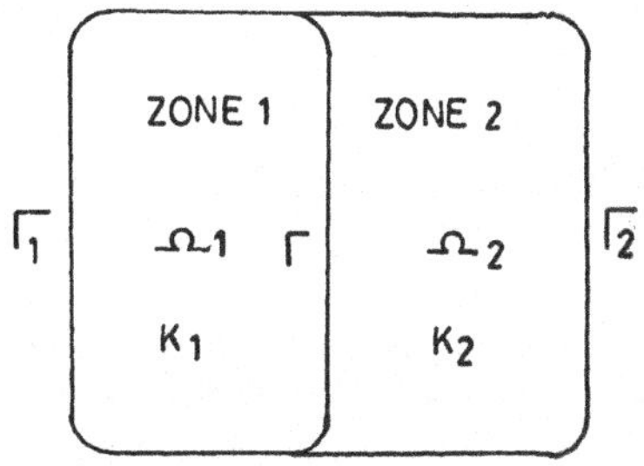

FIG. 1. LAYERED DOMAIN

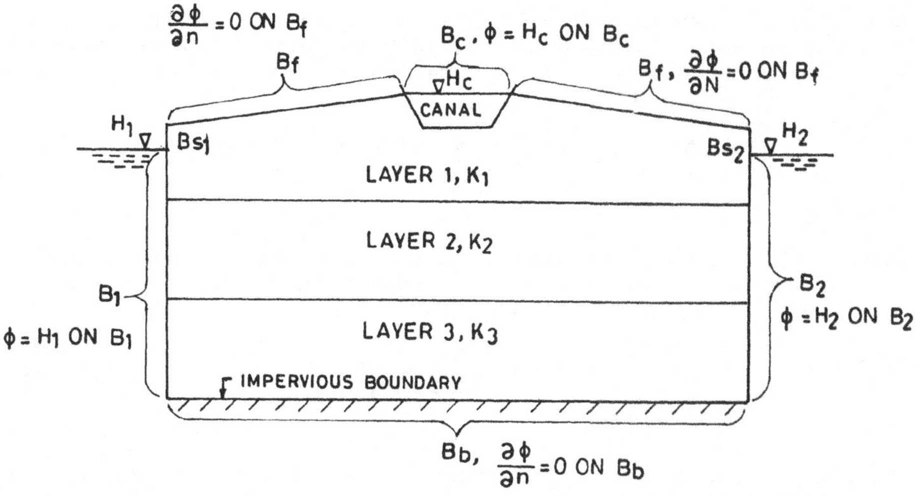

FIG. 2. SCHEMATIC REPRESENTATION OF CANAL – AQUIFER SECTION

layer 1 must be equal to the amount of fluid entering layer 2, that is

$$\left(K_1 \frac{\partial \Phi}{\partial n} \Delta S\right)_{layer\ 1} = -\left(K_2 \frac{\partial \Phi}{\partial n} \Delta S\right)_{layer\ 2} \tag{4}$$

where n is the unit outward normal, s is the sectional distance, K_1, K_2 are the permeabilities of the respective layers.

The boundary conditions to this type of problem are: (1) essential condition such as prescribed head $\phi = \bar{\phi}$ on boundary Γ_1 and (2) a prescribed flux such as $\partial\phi/\partial n = \partial\bar{\phi}/\partial n$ on boundary Γ_2, where n is the unit outward normal on the boundary of the domain Ω. The boundary conditions to this problem are as mentioned in Fig. 2. At the interface between two layers, the compatibility conditions of Equations 3 and 4 have been used.

Considering a two layered medium as shown in Fig. 1, the boundary integral equation for Laplace equation in layer 1 can be obtained by applying Green's theorem on domain Ω_1 [Brebbia 1978]

$$C_1 \Phi_1 = \int_{\Gamma_1} \left(\Phi^* \frac{\partial \Phi_1}{\partial n_1}\right) d\Gamma_1 - \int_{\Gamma_1} \Phi_1 \left(\frac{\partial \Phi^*}{\partial n_1}\right) d\Gamma_1 \tag{5}$$

where ϕ_1 is the potential and n_1 is the unit outward normal on the boundaries of Ω_1, ϕ^* is the fundamental solution of Laplace equation in two dimensions. That is:

$$\Phi^* = \frac{1}{2\pi} \ln\left(\frac{1}{r}\right) \tag{6}$$

where r is the distance from the point of application of the concentrated potential to the point under consideration, c_1 is a constant such that $c_1 = 0$ if the point is outside the domain Ω_1, $c_1 = 1$ if the point is inside the domain and $c_1 = 1/2$ if the point is on a smooth boundary. Similarly, for the domain Ω_2,

$$C_2 \Phi_2 = \int_{\Gamma_2} \left(\Phi^* \frac{\partial \Phi_2}{\partial n_2}\right) d\Gamma_2 - \int_{\Gamma_2} \left(\Phi_2 \frac{\partial \Phi^*}{\partial n_2}\right) d\Gamma_2 \tag{7}$$

in which c_2 and n_2 are similar to c_1 and n_1 .

Multiplying Equation 5 by K_1 and Equation 7 by K_2 gives:

$$C_1 K_1 \Phi_1 + C_2 K_2 \Phi_2 = \int_{\Gamma_1} \left(K_1 \frac{\partial \Phi_1}{\partial n_1} \Phi^* - \Phi_1 K_1 \frac{\partial \Phi^*}{\partial n_1} \right) d\Gamma_1 + \int_{\Gamma_{in}} \left(K_1 \frac{\partial \Phi_1}{\partial n_1} \Phi^* - \Phi_1 K_1 \frac{\partial \Phi^*}{\partial n_1} \right) d\Gamma_{in} + \int_{\Gamma_2} \left(K_2 \frac{\partial \Phi_2}{\partial n_2} \Phi^* - \Phi_2 K_2 \frac{\partial \Phi^*}{\partial n_2} \right) d\Gamma_2 + \int_{\Gamma_{in}} \left(K_2 \frac{\partial \Phi_2}{\partial n_2} \Phi^* - \Phi_2 K_2 \frac{\partial \Phi^*}{\partial n_2} \right) d\Gamma_{in} \quad (8)$$

where Γ_{in} is the interzonal boundary of Ω_1 and Ω_2. Combining the first and third term on the right hand side of Equation 8 and using compatibility conditions of Equations 3 and 4 on the interzonal boundary Γ_{in}, we have:

$$C_i \Phi_i = \int_{\Gamma} \left(\Phi^* K \frac{\partial \Phi}{\partial n} - K \Phi \frac{\partial \Phi^*}{\partial n} \right) d\Gamma + \int_{\Gamma_{in}} \left((K_2 - K_1) \Phi \frac{\partial \Phi^*}{\partial n} \right) d\Gamma in \quad (9)$$

where K is the permeability of the respective zonal boundary, n is the respective outward normal on the respective boundary, $\Gamma = \Gamma_1 + \Gamma_2$ and $C_i = 1$ for any internal point, $C_i = C_1 K_1$ for point on the external boundary of zone 1 (Γ_1), $C_i = C_2 K_2$ for point on external boundary of zone 2 (Γ_2), $C_i = C_1 K_1 + C_2 K_2$ for point on interzonal boundary Γ_{in} and $C_i = 0$ for any external point of the domain Ω.

BOUNDARY DISCRETIZATION:

For a two layered problem, the boundary can be discretised into N elements of which N_{ex} external and N_{in} interzonal elements. The points where the unknown values are considered are called nodes. Here, the concept of constant element is used where the node is taken at the middle of the element and the function values are assumed to be constant over the element. Equation 9 can be rewritten in discretised form as:

$$C_i \Phi_i = \sum_{e=1}^{N_{ex}} \int_{\Gamma_{ex}} \left(\Phi^* K \frac{\partial \Phi}{\partial n} - K \Phi \frac{\partial \Phi^*}{\partial n} \right) d\Gamma - \sum_{e=N_{ex}+1}^{N_{ex}+N_{in}} \int_{\Gamma_{in}} \left((K_1 - K_2) \Phi \frac{\partial \Phi^*}{\partial n} \right) d\Gamma_{in} \quad (10)$$

Adding the contribution of two adjoining elements into one term of Equation 10 and rewriting, we have

$$\begin{aligned} C_i\,\Phi_i = &\sum_{j=1}^{N}\left[\sum_{e=1}^{N_{ex}} g^e_{ij}\right]\frac{\partial\Phi}{\partial n_j} \\ &-\sum_{j=1}^{N}\left[\sum_{e=1}^{N_{ex}} h^e_{ij}\right]\Phi_j \\ &-\sum_{j=1}^{N}\left[\sum_{e=N_{ex}+1}^{N_{ex}+N_{in}} k^e_{ij}\right]\Phi_j \end{aligned} \tag{11}$$

Here, $g_{ij}{}^{e}$, $h_{ij}{}^{e}$, $k_{ij}{}^{e}$ are the influence coefficients defining the interaction between the singular point i and a particular node on an element e, such that:

$$g_{ij}{}^{e} = \int_{\Gamma_e} K\,\Phi^*\,d\Gamma_e \tag{12}$$

$$h_{ij}{}^{e} = \int_{\Gamma_e} K\,\frac{\partial\Phi^*}{\partial n}\,d\Gamma_e \tag{13}$$

$$k_{ij}{}^{e} = \int_{\Gamma_{in}} (K_1 - K_2)\,\frac{\partial\Phi^*}{\partial n}\,d\Gamma_{in} \tag{14}$$

Combining $h_{ij}{}^{e}$ and $k_{ij}{}^{e}$ terms, we have:

$$\begin{aligned} \overline{H_{ij}} &= \sum_{e=1}^{N_{ex}+N_{in}} h_{ij}{}^{e} \\ \overline{G_{ij}} &= \sum_{e=1}^{N_{ex}} g_{ij}{}^{e} \end{aligned} \tag{15}$$

Equation 11 becomes:

$$C_i\,\Phi_i = \sum_{j=1}^{N}\overline{G_{ij}}\,\frac{\partial\Phi_j}{\partial n} - \sum_{j=1}^{N}\overline{H_{ij}}\,\Phi_j \tag{16}$$

Equation 16 can be written as:

$$H\,\Phi = G\,\frac{\partial \Phi}{\partial n}$$

$$\text{where} \quad H_{ij} = \sum_{e=1}^{N} h_{ij}^{e}\,, \quad \text{if } i \neq j \tag{17}$$

$$\text{and} \quad H_{ij} = \sum_{e=1}^{N} h_{ij}^{e} + C_i\,, \quad \text{if } i = j$$

After applying boundary conditions, Equation 17 can be expressed as:

$$A\,X = F \tag{18}$$

where X is a vector of unknowns in ϕ and $(\partial\phi/\partial n)$, F is a known matrix of size N X 1, A is the coefficient matrix of size N X N.

SEEPAGE CALCULATIONS:

After obtaining the solution on the boundary, the seepage can be calculated from the given boundary conditions and permeability of the medium. On the canal section, ϕ is known and hence $\partial \phi/\partial n$ can be obtained. The seepage quantity from the canal is given by

$$Q_s = \int_{\Gamma} K\,\frac{\partial \Phi}{\partial n}\,d\Gamma \tag{19}$$

where Γ is the wetted perimeter of the canal.

The presence of free surface produces a system of nonlinear equations which are to be solved by an iterative procedure. This procedure is based on the assumption that the free surface is a no flow boundary and the head at each point on the free surface is equal to its elevation. The procedure is as follows:

1. Assume an initial free surface position, y=f(x).
2. Discretize the domain and determine the potential at the free surface ϕ = g(x), such that $\partial\phi/\partial n = 0$.
3. Compare f(x) and g(x). If the difference is within a permissible tolerance limit, y = g(x).
4. If the difference is large, another iteration with y - coordinate of free surface node as y = 0.5(f(x) + g(x)).
5. This procedure is repeated till convergence is achieved.

CASE STUDY:

As a case study, a canal over a 3 layered porous medium as shown in Fig. 3 has been chosen. The water table on the left and right is assumed to be at 50 m from the impervious bottom of the section. The head in the canal is assumed to 60 m. The extent of the medium chosen for analysis is 150 m on either side of the center-line of the canal. The permeability values chosen for top, middle and bottom layers are 1.25, 2.5 and 5.0 m/day respectively.
The free surface of the water table has been adjusted in such a way that flux normal to the free surface is zero, using the iterative procedure described above. The discretization of the boundary using constant boundary elements is as shown in Fig. 3'.

As the domain under study is symmetric about the center-line of the canal section, it is sufficient if half the domain is analysed. The left hand side of the section has been divided into 53 boundary elements. The seepage flux from the canal varies from 0.005478 on the side to 0.39849 on the bottom. The seepage for each element was calculated and added. The total seepage per meter length of the canal is 2(1.25)[0.2425(2) + 0.39849(2) + 0.3176 + 0.1541(2.828) + 0.05478(2.828) + 0.135017(1.414)] = 5.602 m^2/day/m.

As a comparative study, the same problem was analysed using Finite Element method which was validated with an electrical analogue model [10]. The Finite Element discretization of the problem is as shown in Fig. 4. Due to the symmetry, half section of the domain is analysed by taking 36 linear triangular elements and 30 nodes. The seepage value obtained is 5.574 m^2/day/m which is very close to the value obtained by boundary element method.

CONCLUSIONS:

This study shows that the boundary element method can be effectively used to solve steady state seepage flow from canals to the adjacent layered porous media. The specific advantages of the boundary element method like reduction in dimensionality, ease with which data can be handled, reduced size of the matrix and ease with which the free surface problems can be handled, simplifies the problems to be handled compared to the other numerical methods. Eventhough singularity appears at the corners of the domain and at intersection corners of layers, a reasonably satisfactory result is obtained after avoiding the singularity in the formulation. With minor modification, the steady state analysis can be extended to unsteady seepage analysis also.

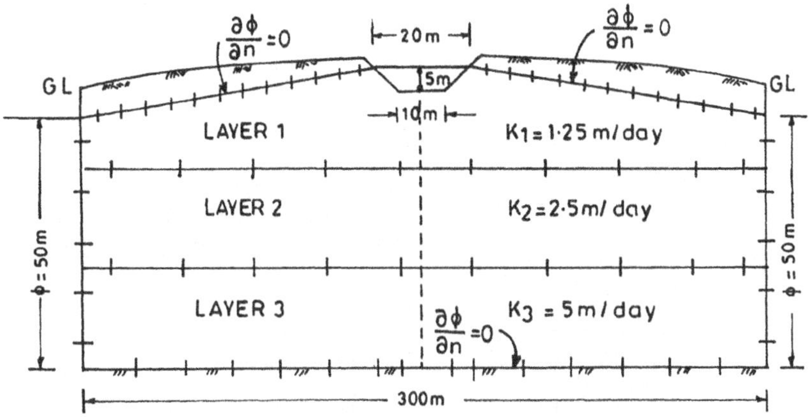

FIG. 3. BEM DISCRETIZATION OF CANAL SEEPAGE PROBLEM

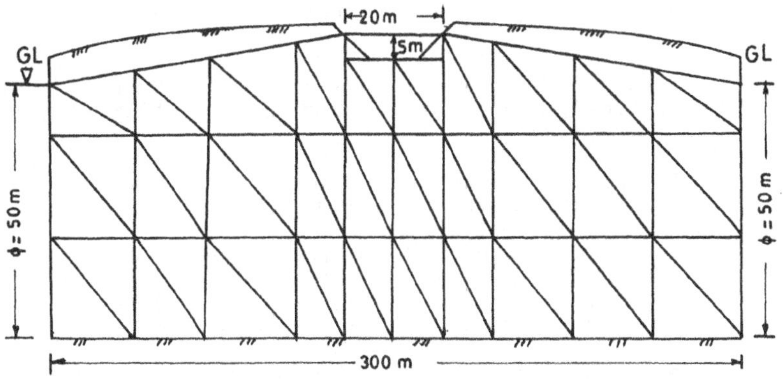

FIG. 4. FEM DISCRETIZATION OF CANAL SEEPAGE PROBLEM

REFERENCES:

1. Bear Jacob, "Hydraulics of Ground Water", McGraw-Hill Book Co., New York, 1979.

2. Brebbia, C.A. "Boundary Element Method for Engineers' Pentech Press Ltd., London, 1978.

3. Bouwer, H. 'Theory of Seepage from Open Channels' in 'Advances in Hydrosciences' Edited by Ven Te Chow, Vol. 5, Academic Press, New York, pp. 121 - 172, 1969.

4. Eldho, T.I. "Boundary Element Analysis of Fluid Flow Problems",M.Tech. dissertation, Department of Civil Engineering, Indian Institute of Technology, Bombay, 1992.

5. Garg, S.P. and Chawla A.S. ' Seepage from Trapezoidal Channels', Journal of Hydraulics Division, ASCE, 96, HY6, pp.1261-1282, 1970.

6. Huebner K.H. and Thornton E.A. 'The Finite Element Method for Engineers', John Wiley and Sons Inc., New York, 1982.

7. Jia Dong Zheng 'LOcation of Free surface for Zoned Problems A New Economical Boundary Element Method', Proc. of 5th International Conference on Boundary Elements, Edited by C.A.Brebbia, Japan, 1983.

8. Ligget, J.A. and Philip L.F. 'The Boundary Integral Method for Porous Media Flow', George Allen and Lewis Publishers, London, 1983.

9. P. Ya Polubarinova - Kochina 'Theory of Ground Water Movement', Princeton University Press, New Jersey, U.S.A., 1962.

10. Rastogi, A.K. and Joshi S.G. 'Digital Modelling to Investigate Canal Seepage Losses', Research Report submitted to Dept. of Science and Technology, Govt. of India, Department of Civil Engineering, Indian Institute of Technology, Bombay, 1991.

11. Vasudeva Rao B. and Eldho T.I. "Estimation of Canal Seepage by Boundary Element Method", Proceedings of the Eigth Asia Pacific Division of IAHR Conference, Pune, India, 1992.

SECTION 3:

GROUNDWATER RECHARGE

GROUNDWATER - SURFACEWATER INTERMIXING MODEL AND RECHARGE CONDITIONS IN DELHI AREA AS DERIVED FROM $\delta^{18}O$ AND δD

P.S.DATTA AND S.K.TYAGI
Nuclear Research Laboratory
Indian Agricultural Research Institute
NEW DELHI - 110012, INDIA

ABSTRACT

The data on oxygen-18 and deuterium content in Delhi groundwater show a wide range and indicate that a selection effect in favour of water depleted in stable isotope plays an important role in recharging groundwater adjacent to canal and river. Wide range of stable isotope composition of rainwater in the area and inhomogeneity of the rainfall recharge to the groundwater, as well as mixing of water through different flow paths result in spatial variation in stable isotope content of groundwater. By adopting a very simplified model of equal inflow through the screens to the well with depth, the isotopic gradient with depth has been determined. This simple model on depth variation of stable isotopic content in groundwater indicates contribution of canal water and river water to the groundwater to a considerable depth of the aquifer system. There is a large variation in salinity of groundwater. There is an isotope and chloride content gradient in the upper shallow groundwater, seem to be caused due to diffusion of water from shallow aquifers.

INTRODUCTION

In meteoric water of moderate climates the isotopic composition of D and ^{18}O are related by the equation $\delta D = 8\ \delta^{18}O + d$ (Craig, 1961) where δ - values are expressed in parts per thousand relative to SMOW (Standard Mean Ocean Water). When plotted on linear ($\delta^{18}O$, δD) diagram the line is referred to as the meteoric line. Isotopic composition of waters are generally interpreted by use of such a diagram. On the premise that shallow groundwater represents an 'averaged contemporary rainfall' the isotopic ratios in groundwaters can be considered as a representative for isotopic composition of rainfall in a region. Since isotopic exchange with soil is a very slow process at normal temperatures, practically no exchange takes place during the course of groundwater recharge.

In recent years efforts have been devoted to studies of the $^{18}O/^{16}O$ and D/H ratios in groundwaters and rainfall on the Indian Subcontinent.The measurements of the isotopic ratios in groundwaters in India (Bhattacharya et al., 1985) indicate a continental effect in the case of shallow groundwaters. The inland gradient in $\delta^{18}O$ is found to be -2 ‰ per 1000 km between the

V. P. Singh and B. Kumar (eds.), Subsurface-Water Hydrology, 103–117.

coastal Calcutta and inland Delhi (East-Northwest). However, this $\delta^{18}O$ change with respect to distance is much less than the gradient of -6 ‰ per 1000 km seen for the shallow groundwaters along a traverse from Ahmedabad to Delhi (West-Northwest). The study also shows the unambiguous effect of isotopic enrichment due to evaporation from soils.

The studies in Delhi region (Das et al. 1988) show wide variations in the isotopic composition of groundwater with $\delta^{18}O$ ranging from -4.18 ‰ to -7.5 ‰and δD ranging from -36.8 ‰ to -54.0 ‰. As compared to this, the yearly weighted mean $\delta^{18}O$ values of rainfall (1961-82) in New Delhi ranges from -4.18 ‰ to -8.37 ‰ and δD ranges from -24.2 ‰ to -60.5 ‰ (IAEA, 1981). However, the δ-values of monthly rainfall (1961-83) at New Delhi shows a much wider range during June to September with $\delta^{18}O$ values from +4.9 ‰ to -15.3 ‰ and δD varying from + 56 ‰to -120 ‰. In Delhi region, the $\delta^{18}O$ and δD composition of rainfall vary with time due to changing meteorological conditions such as rainfall intensity and distribution, temperature, rainy spells, evaporation of rain drops during its fall, etc. (Datta et al, 1991). Very high standard deviation of δ-values for each month (IAEA, 1981), particularly for winter months (November to February) and Summer months (March-June) makes it difficult to estimate a reliable rainfall input function. The present paper describes the reasons for wide variation in isotopic composition of groundwater and important features of groundwater recharge and groundwater-surfacewater intermixing in the western and southern parts of Delhi.

DESCRIPTION OF THE AREA

The Delhi area is a part of the Indo-Gangetic alluvial deposits and is at an elevation ranging from 198 to 220 m. The area is transacted by rocky ridge of quartzite belonging to the Aravalli system. The alluvial part of the plain also contains some old sand dunes, while the rest is water transported material of different ages.The union territory of Delhi with an area of 1485 sq. km lies between 28°24'17"- 28°53'00" N and 76°50'24" - 77°20'37" E. Two-third of the area under irrigation is irrigated by groundwater extensively. North-south subsurface geology indicate predominance of clay and silt intermixed with fine sand. Kanker layer occur at a depth of 2.4 to 3.0 meters below the surface. During the hottest months of May and June temperature commonly exceeds 40°C. Average annual rainfall in the area is 53 cm.

Groundwater occurs under the following three hydrogeological conditions :

1) Alluvial basin of Chhattarpur in the south of Delhi, enclosed within rocky surroundings of the Delhi ridge. The basin acts as a single aquifer under unconfined groundwater conditions.

2) Alluvial deposits to the west of Delhi ridge. The aquifer is under semi-confined conditions. and

3) Alluvial deposits to the east of the ridge :

(a) between the ridge and the Yamuna river and (b) east of Yamuna

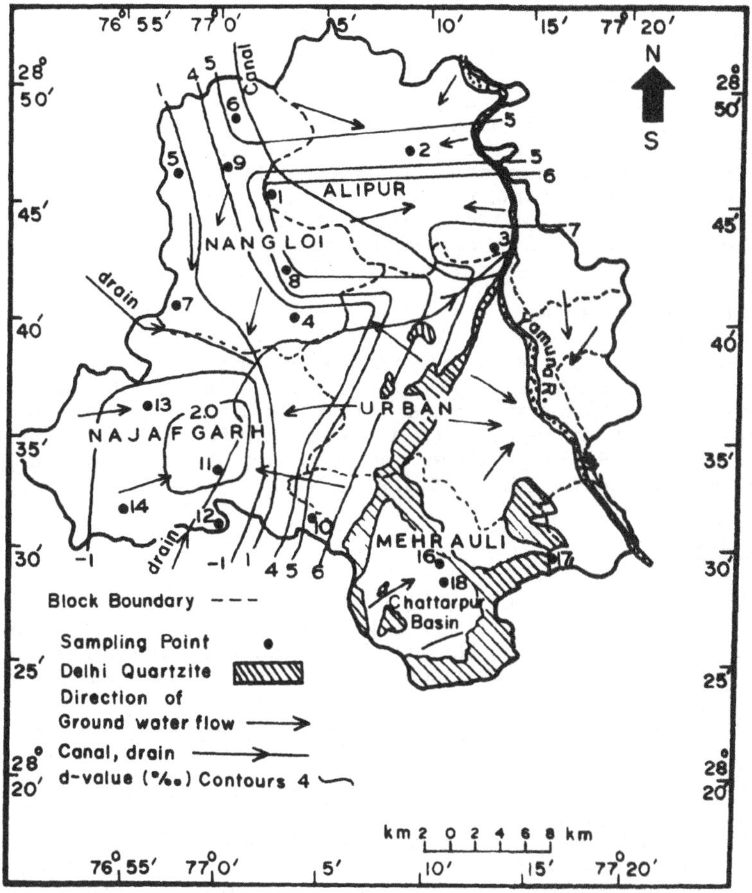

Fig.1. Map of Union Territory of Delhi showing location of sampling points and deuterium excess (‰) isocontours.

river. The aquifers are under semi-confined conditions.

The Delhi ridge acts as a groundwater divide between the western and eastern parts of Delhi. The water table slopes from the ridge on both sides.The water table is relatively shallower in northern and north-western parts and deeper levels exist in the southern parts. South-western parts of the area has a large groundwater trough at Najafgarh block. To the west of the region, groundwater drains towards this trough. To the east, groundwater discharges to the Yamuna river. Owing to the relatively low permeability of the water bearing formations, the movement of groundwater is generally slow and sluggish.

MATERIALS AND METHODS

The data on stable isotopic composition and chloride content of groundwaters (Das et al, 1988) are given in Table 1. The analytical accuracy is ± 0.15 ‰ for $\delta\ ^{18}O$ and ±1 ‰ for deuterium. The δ-values are expressed with respect to isotopic ratio of V-SMOW standard (Gonfintini, 1978).Fig.1 shows the location of the sampling points and deuterium excess (d) contours.

DISCUSSION

δd - $\delta^{18}O$ relationship of Delhi region groundwater has been shown in Fig.2. Wide range and spatial inhomogeneity in the isotopic composition indicate that the aquifer in the western parts of Delhi does not constitute a homogenous system in its lateral extent.

In Fig.2 the points show a clear deviation from the Delhi Meteoric Line (1961-82) as computed from the IAEA isotope data on Delhi rainfall (IAEA, 1981). A closer look at the data in Fig.2 indicates that except samples no.11, 12,13 and 14 of Najafgarh block all other points fall within the standard deviation of the slope of the meteoric line. Slope less than 8 indicates that local effects play an important role in controlling isotopic composition of the groundwater in the studied area. One possible reason may be that poor infiltration properties of the soil (due to high clay content)results in isotopic enrichment of water evaporated from soil before recharge takes place. However, normally such areas contribute very little to the recharge of groundwaters of their vicinity, because most of the rainfall flows as surface run-off.

The total weighted mean of the yearly weighted mean δ-values of Delhi rainfall (1961-82) has been computed to be -6.09 ‰ for $\delta^{18}O$ and -44.0 ‰ for δD. The average isotopic composition of the groundwater (samples 16,17 and 18) in the closed basin of Chhattarpur having unconfined aquifer works out to be -6.09 ‰ for $\delta^{18}O$ and -46.1 ‰ for δD. The mean isotopic composition of the groundwater in western parts of the Delhi region, having semi-confined aquifer system, is -6.10 ‰ for $\delta^{18}O$ and -46.7 ‰ for δD. These δ-values are well in agreement with yearly weighted mean δ-

Table 1. Stable isotopic composition and chloride content of groundwater in Delhi region.

Sample No.	Location	Depth range (m)	Chloride (mg/l)	δD (‰)	$\delta^{18}O$ (‰)	Deuterium Excess (‰)
1.	Pooth Khurd	23.78-29.57 39.60-41.46	220	-53.6	-7.46	6.08
2.	Pehlad Pur	21.95-27.44 44.82-50.00	465	-51.1	-6.99	4.82
3.	Wazirabad	9.15	237	-53.4	-7.57	7.16
4.	Nangloi	13.72	889	-45.4	-6.01	2.68
5.	Punjab Khor	31.40-49.70	319	-45.5	-5.71	0.18
6.	Daryapur Kalan	33.54-58.50 64.00-69.50	359	-54.0	-7.47	5.76
7.	Tikri Kalan	9.15	1703	-46.4	-5.81	0.08
8.	Nihari	12.20	1709	-49.4	-6.87	5.56
9.	Chandpur	9.15	5240	-48.9	-6.63	4.14
10.	Kapas Hera	106.7	11	-48.3	-6.67	5.06
11.	Chhawla	31.49	117	-36.8	-4.18	-3.36
12.	Raghupur	18.29	1885	-44.1	-5.30	-1.70
13.	Mitraon	50.61	618	-41.9	-5.04	-1.58
14.	Guman Hera	12.20	1750	-38.2	-4.61	-1.32
15.	Bapania	14.63	2875	-44.9	-5.60	-0.10
16.	Chhattarpur	12.20-50.30	133	-44.2	-5.72	1.56
17.	Madan Gir	65.40-74.70	149	-46.7	-5.94	0.82
18.	Sat Beri	18.90-73.17	51	-47.5	-6.37	3.46

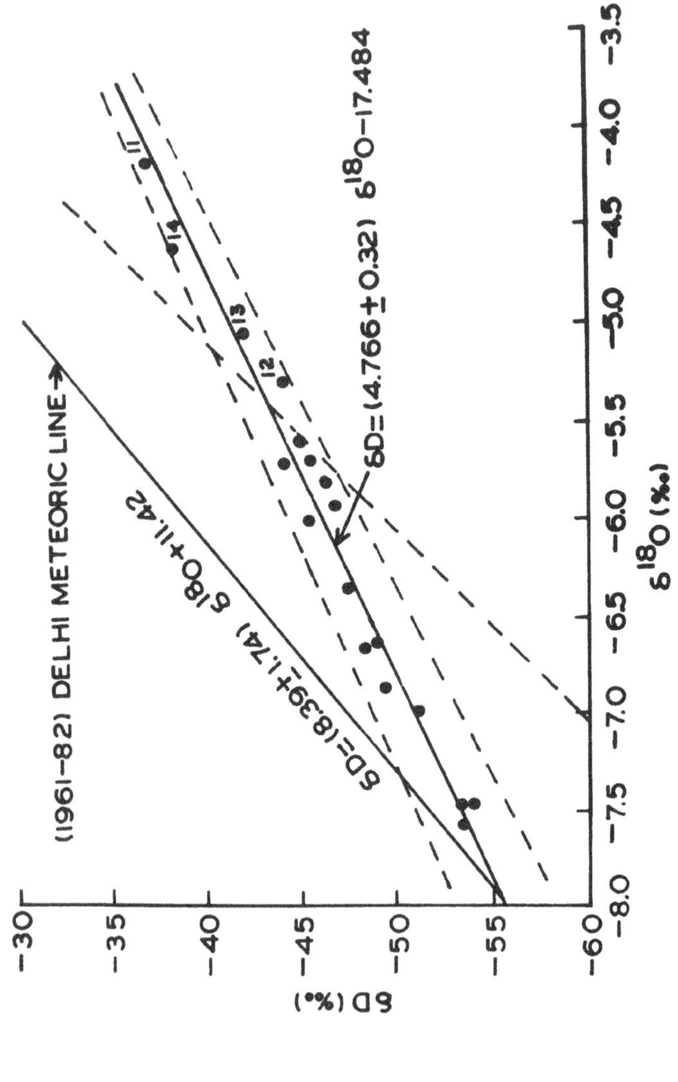

Fig. 2. $\delta D-\delta^{18}O$ Relationship of Delhi Ground Water. Dotted lines indicate standard errors of the slopes of the regression lines.

values. Thus the mean isotopic composition of groundwater in the region can be considered to be the recharge index by infiltration of yearly rainfall (Datta, 1990). However, It is seen from Fig.2 that the groundwater samples west of ridge in Najafgarh block (samples no. 11,12,13 and 14) show enrichment with reference to the yearly weighted mean values. While the groundwater samples of other areas either depleted or fall within ± 0.5 ‰ of the yearly weighted mean $\delta^{18}O$. The wells having isotopically depleted groundwater (samples no. 1,2,3,6,8 and 9) are located near the canal and river.

Since the rainfall during the monsoon months (July, August and September) mainly contributes to groundwater recharge in the region, it would be more realistic to consider the mean δ-values of rainfall in the monsoon months as the recharge index. The mean δ-values of the monsoon months rain are -7.23 ‰ for $\delta^{18}O$ and -47.8 ‰ for δD. It is seen that except for the samples (No.1, 2,3,6,8 and 9) near the river or unlined canal, which fall within ±0.5 ‰ around the δ-values of the monsoon months rainfall, most of the other groundwater samples are isotopically enriched than the δ-values of the monsoon months rain. This does not seem to be due to evaporation of groundwater, because the groundwater in deeper aquifers (more than 20 m depth) is not likely to be affected by direct evaporation. However, the possibility of the isotopic enrichment due to evaporation of water from shallow aquifers (<20 m depth) can not be ruled out, which has been discussed separately. Therefore, a selection effect in favour of depleted waters appears to be occurring in the groundwater recharge process only in aquifers adjacent to the canal and river.

Distortions in stable isotopic composition during recharge of rainfall is known to be relatively small and rarely exceeds ± 1 ‰ in $\delta^{18}O$ (Gat, 1983). Hence, spatial variability and non-steady nature of recharge from rainfall having different δ-values and groundwater exploitation are the possible reasons which results in mixing of waters from different flow-paths, causing variation in isotopic composition of groundwater in Delhi region.

Depth variation of $\delta^{18}O$ and δD in groundwater

Variation of $\delta^{18}O$ and δD in groundwater with depth range of aquifer has been shown in Fig.3 and Fig.4. The variability can be grouped under two distinct geohydrological situations, viz., (1) Aquifers under semiconfined and unconfined conditions to the west of ridge and Chhattarpur basin, and (2) Aquifers adjacent to the river and unlined canals and drains. While under situation (1) there is an apparent trend of depletion in δ-values with increase in depth, under situation (2) there seem to be no depth variation trend. The possible trends of depth variation under situation (1) have been shown by solid line in Fig.3 and Fig.4. The data shows a possible existence of stratification of groundwater deeper than 20 m depth under situation (1). In the case of wells having screen, average depth has been taken into consideration.

The drilled wells have screens extending over the major depth of the wells. Therefore the collected water samples integrate the

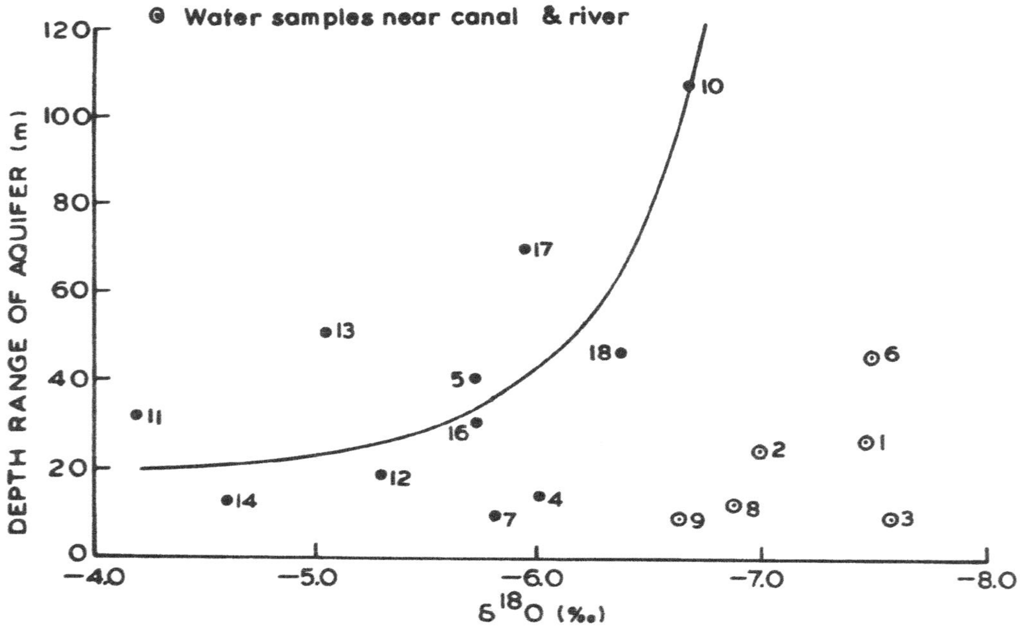

Fig. 3. Variation of $\delta^{18}O$ with depth in Delhi ground water to the west of ridge and chattarpur basin. Possible trend of variation has been shown by solid curve.

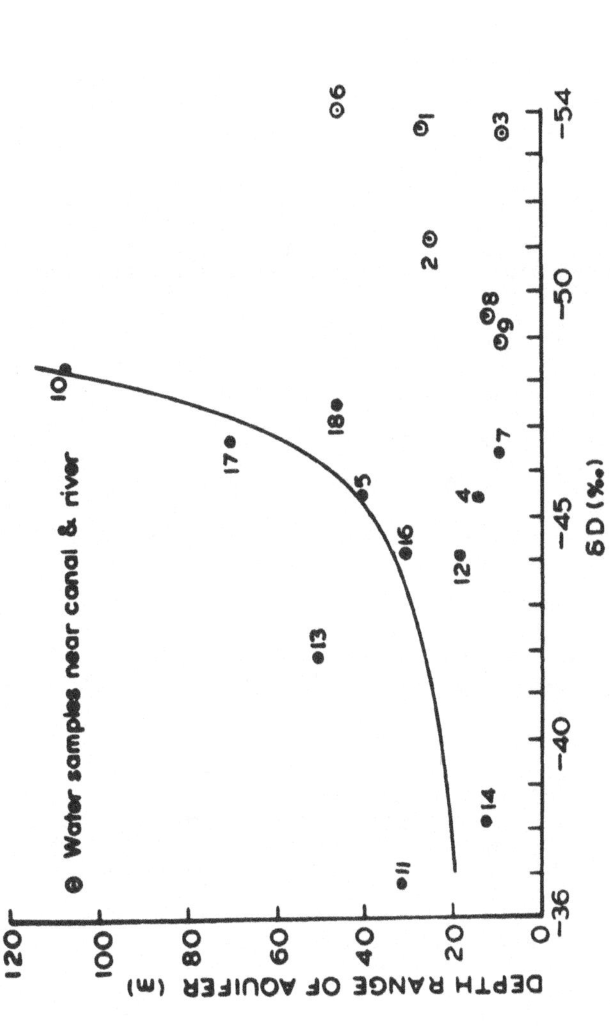

Fig. 4. Variation of δD with depth in Delhi ground water to the west of ridge and chattarpur basin. Possible trend of variation has been shown by solid curve.

isotopic composition of groundwater down to the depth upto which the screens exist. By adopting a very much simplified model of equal inflow through the screens to the well with depth, the isotopic gradient with depth can be determined. For the linear model, the observed isotopic composition of groundwater (δ_o) can be related to the gradient (b) and the limits (x_1 , x_2) of the screens by the following equation :

$$\delta_o = \delta_s + (x_1 + x_2)/2 \qquad \text{.............Eqn. (1)}$$

Where δ_s is the isotopic composition of the shallow groundwater.

In the case of wells no. 1,2 and 6, the maximum depth upto the screens exist are 30 m, 50 m and 70 m respectively (Table 1). By adopting the above model for well no. (1) and (2) the following two equations are obtained.

Well No. (1) $-7.46 = \delta_s + b/2\ (5.79 + 1.86)$ .....Eqn. (2)

Well No. (2) $-6.99 = \delta_s + b/2\ (5.49 + 5.18)$ .....Eqn. (3)

by solving these two equations :

$b = 0.3112 \qquad \delta_s = -8.65$

Therefore, δ-value at depth x can be expressed by the following equation :

$$\delta_x = -8.65 + 0.3112\ x \qquad \text{.........Eqn. (4)}$$

Similarly by considering the well no. 1 and 6, the equation for depth variation of δ-values works out as follows :

$$\delta_x = -7.43 - 0.0087\ x \qquad \text{..........Eqn. (5)}$$

The well no.3 being located very near to river, the δ-values of sample no.3 ($\delta^{18}O$ = -7.57 ‰) can be considered to be the river index. The well no. 1 and 6 being located adjacent to canal, the mean δ-values of these samples (mean $\delta^{18}O$ = -7.47 ‰) can be considered to be the canal index, which is almost same as the river index.

Groundwater-Surfacewater Intermixing

The equation (4) shows isotopic enrichment with increase in depth. The δ-value at 3 m depth works out to be -7.71 ‰ which is similar to the river index. Therefore, in the case of well no.2 and the area surrounding it in Alipur block, there is decreasing importance of river and canal recharge component with depth. Beyond 5 m depth upto 10 m depth the groundwater is originated from the rainfall recharge, because the δ-value computed is very near to the rainfall mean δ-value. In the same way, equation (5) shows depletion with depth, although with small depth variation. At 10 m depth δ-value is -7.51 ‰, which is similar to the canal index. The δ-value does not change very much upto 30 m depth. Thus it appears that in the case

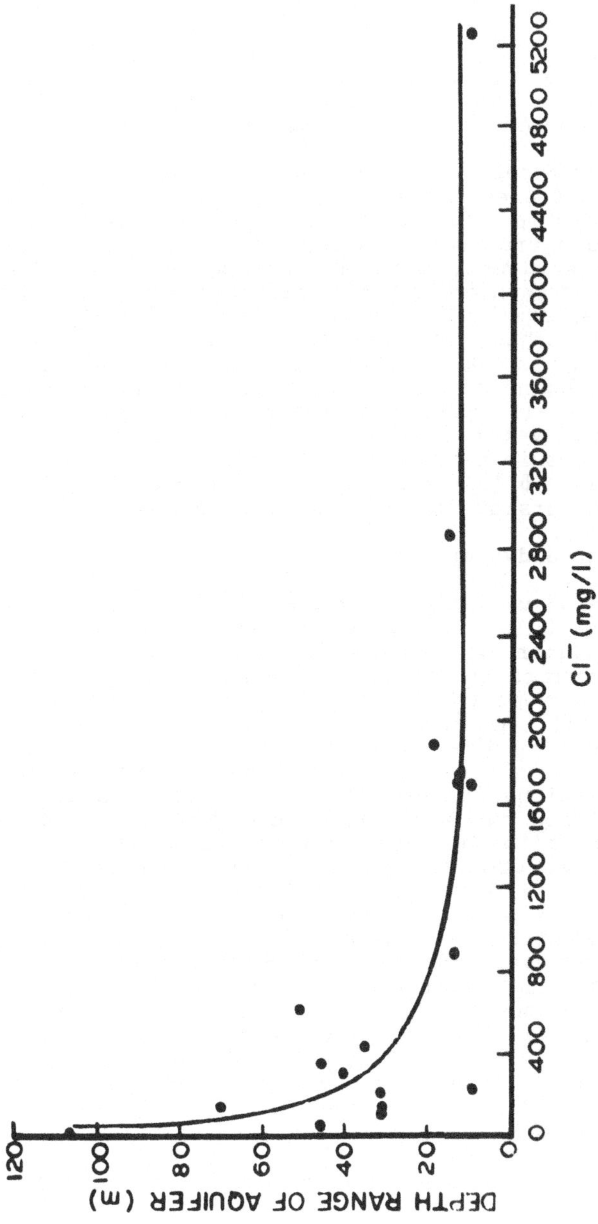

Fig. 5. Trend of chloride content in Delhi groundwater with depth range of aquifer.

of well no. 1,6,8 and 9 the canal water contributes to the groundwater recharge to a considerable depth. It is evident that regular analyses of stable isotope composition of rainfall, groundwater, canal water and river water can provide quantitative estimates of rainfall recharge,canal contribution and river contribution components in groundwater.

Deuterium excess (d) values, with a wide range from -3.36 ‰ to +7.16 ‰ for different water samples are given in Table -1. Iso-contours of d-values in Fig.1 indicate that the deuterium excess values decrease from north to south-west. d-values significantly less than 10 ‰ in most cases may not be due to only evaporation effect, because such effect could be prominent in shallow aquifers and not in deep aquifers. Less d-values and slope less than 8 in δD - $\delta^{18}O$ relationship can also result from mixing of waters of different d-values. Delhi rainfall also show a wider range of d-values which depend on the intensity and distribution of rainfall (Datta at el, 1991), that contributes to the groundwater recharge. Mixing of groundwaters of different d-values is evident form the linear relationship between 'd' and $\delta^{18}O$.

$$d = -3.234 \;\; \delta^{18}O - 17.485 \qquad \text{Corr. coeff.} = 0.95$$

such mixing occurs due to pumping of groundwater.

Salinity in groundwater

There is a large variation in salinity of groundwater in the area to the west of Delhi ridge. In general, at shallow levels (<20 m depth) groundwater is moderately to highly saline in the western and southwestern parts of Delhi. At deeper levels also groundwater is saline in many parts. The possible trend of chloride content in groundwater with mean depth range of the aquifer is shown by solid curve in Fig.5. The data indicate that diffusion of shallow aquifer water salinity into the deeper waterbody may be responsible for the observed variations in chloride content with depth. The shape of the concentration curve in the figure shows that real system may behave according to the second law of diffusion.

The shallow groundwaters to the west of ridge show a good correlation between $\delta^{18}O$ and chloride content (Fig.6). A hypothesis that might explain such a correlation is that an isotope and chloride concentration gradient exists in the upper shallow groundwater because of evaporation. The thickness of such a zone is probably rather limited and presumably depends on the depth of the water table. No appreciable change in isotopic composition with change in chloride content (samples no. 8 and 9) indicates the possibilities of salinization due to nutrient leaching from agricultural activities in the west and north-west parts of Delhi.

CONCLUDING REMARKS

The groundwater to the west of ridge and in northern parts of Delhi is replenished by different sources such as rainfall, canals and river, each having its own stable isotopic composition.A selection

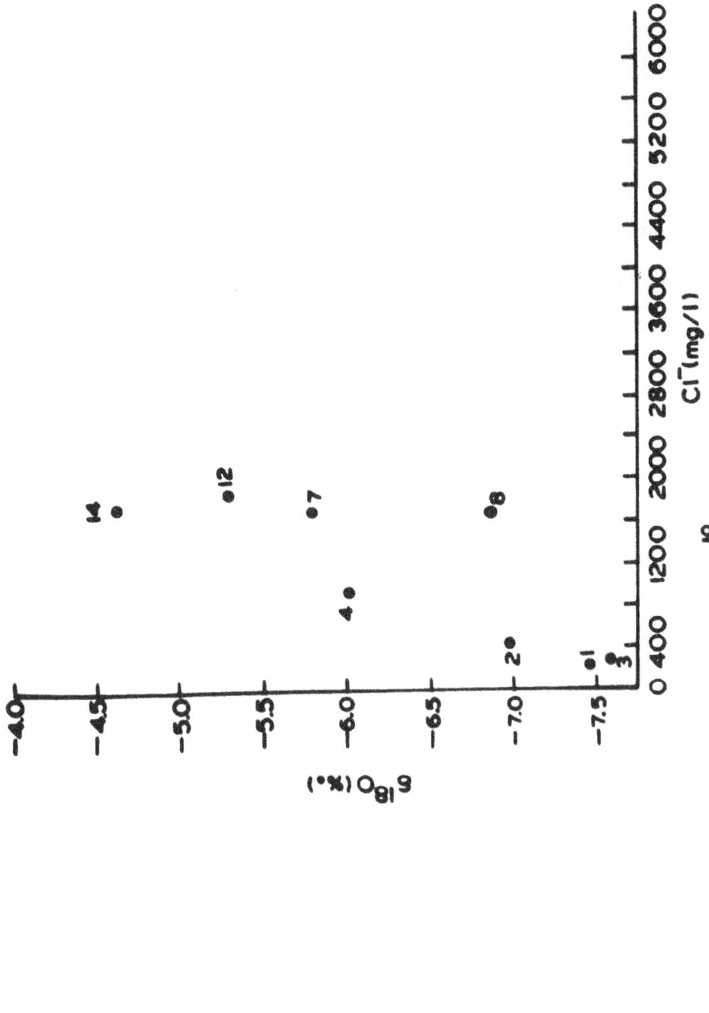

Fig.6. Correlation of $\delta^{18}O$ and chloride content in shallow ground waters (depth<20m) of Delhi to the west of ridge, indicating evaporation effect.

effect in favour of water depleted in stable isotopes appears to play an major role in recharging the groundwater adjacent to the canal and river. Variations in the isotopic composition of the groundwater in lateral extent is caused by inhomogeneity of rainfall in different parts and due to variability of isotopic composition of the rainfall input to the groundwater system.

Non-steady flow and mixing pattern of the groundwater, caused either by variations in the amount of input water or by over-exploitation of aquifers by pumping, can transform steady spatial variation of the isotopic composition and salinity of groundwater into changes with time. Evidently, the stable isotopes technique can provide useful information on quantitative estimates of the water sources contributing to the groundwater recharge. Further investigation on the seasonal variation of stable isotopes in rainfall, groundwater and surface water can significantly improve our knowledge of the problem.

ACKNOWLEDGEMENT

The authors thankfully acknowledge Dr. P.N.Tiwari, Project Director, Nuclear Research Laboratory, Indian Agricultural Research Institute, New Delhi for his encouragement in isotope hydrology studies.

REFERENCES

Bhattacharya, S.K., Gupta, S.K., and Krishnamurthy, R.V., 1985. 'Oxygen and hydrogen isotopic ratios in groundwaters and river waters from India'.*Proc. Indian Acad. Sci. (Earth Planet Sci.)*, **Vol.94**, No.3, pp.283-295.

Craig, H., 1961. 'Isotopic variations in Meteoric waters'. *Science*, **Vol.133**, pp.1702-1703.

Das, B.K., Kakar, Y.P., Moser, H. and Stichler, W., 1988. 'Deuterium and Oxygen-18 studies in groundwater of the Delhi area, India'. *J. Hydrology*,**Vol.98**, pp.133-146.

Datta, P.S., 1990.'Deuterium and Oxygen-18 studies in groundwater of the Delhi area, India-comment'. *J. Hydrology*, **Vol.113**, pp.385-389.

Datta, P.S., Tyagi, S.K. and Chandrasekharan, H.,1991.'Factors controlling stable isotope composition of rainfall in New Delhi, India'. *J. Hydrology*, **Vol.128**, pp.223-236.

Gat, J.R., 1983. 'Precipitation, groundwater and surface waters. Palaeoclimates and palaeowaters - a collection of environmental isotope studies'. *Proc. Advisory Group Meeting, Vienna, 1980, IAEA Report*, pp.1-12.

Gonfiantini, R., 1978. 'Standards for stable isotope measurements in natural compounds'.*Nature*, **Vol.271(5645)**, pp.534-536.

IAEA, 1981.'Statistical treatment of environmental isotope data in precipitation'. *IAEA Tech. Rep. Series 206, Vienna*.

EVALUATION OF RECHARGE TO GROUNDWATER DUE TO APPLIED IRRIGATION UNDER DIFFERENT CROPS IN UTTAR PRADESH USING ISOTOPE TRACER TECHNIQUE - A FIELD STUDY.

P.K.Arora
Director
U.P.Irrigation Research Institute, Roorkee-247667

S.N.Rawat, S.S.Chhabra and Suraj Prakash
Ground Water Division
U.P.Irrigation Research Institute, Roorkee-247667

ABSTRACT. A field study using Isotope tracer technique has been carried out to evaluate recharge to groundwater due to applied irrigation under different crops viz. wheat, paddy and sugarcane in three canal command areas of Uttar Pradesh having different hydrogeological conditions. As a result of this study the average recharge to groundwater due to applied irrigation has been worked out as 25 to 32%, 51 to 59% and 27 to 40% for wheat, paddy and sugarcane respectively. The study indicates that Isotope tracer technique provides better alternate to evaluate recharge due to applied irrigation under different crops and consequently results in better planning of optimum use of water resources.

INTRODUCTION

India having a predominantly agricultural economy needs a planned and optimal utilization of all water resources to meet the rapidly increasing demand of water for irrigation. In this context, integrated development of groundwater which forms the chief component of water resources attains an over-riding importance. In fact, the prosperity of the country would depend largely on the extent to which this replenishable resources can be utilized judiciously.

Evaluation of groundwater recharge from various sources forms an important aspect of groundwater resource assessment. Recharge due to applied irrigation both from groundwater and surface water sources is one of the parameter and its quantification by actual field observation is of paramount importance for drawing up plans for the utilization, management and observation of avaialable water resources. The water application has to meet crop consumptive use by

V. P. Singh and B. Kumar (eds.), Subsurface-Water Hydrology, 119–130.

potential evapo-transpiration and losses due to percolation. The loss of percolation is largely a function of the assortment, organic contents, swelling properties of clays, root penetration and puddling etc. Recharge due to percolation is quite large in States where paddy and sugarcane is predominent.

Conventional methods used for estimating recharge due to applied irrigation comprises of (i) Water balance approach and (ii) Drum culture technique or lysimeters. Even the use of groundwater balance approach is often restricted on account of the non-availability of precise data on rainfall, runoff, specific yield, evapo-transpiration and water table fluctuations etc. Lysimeter or drum culture technique, however, is very expensive, and further it is not possible to construct and maintain a large number of lysimeters at more number of places in a basin to get representative data for the entire basin. The benefit of the tracer technique is that studies generally conducted in the field and vertical movement of irrigation water below root zone can be traced directly and no field data as mentioned above is required.

The present paper deals with the field studies carried out by Isotope Group of U.P.Irrigation Research Institute, Roorkee using Isotope tracer technique, to evaluate percentage of applied irrigation recharging the groundwater table under different crops viz. wheat, paddy and sugarcane in three different canal command areas having different hydrogeological conditions. The study is based on the data collected for a period from 1985 to 1986.

AREA AND ITS GEOHYDROLOGY

Recharge due to applied irrigation has been carried out in three canal commands viz. Deoband branch, Sarda Sahayak canal and Saryu canal of Western, Central and Eastern Uttar Pradesh under different crops having different hydrogeological conditions. The locations of three areas are shown in Fig. 1. However, in the present paper one case study viz. Deoband branch command area has been described in detail while the results of the other command areas have been reported.

Deoband branch command area comprising partly the districts Saharanpur and Muzaffarnagar lies between latitudes 29°4'54" to 30°0'0" N and longitudes 77°14'17" to 77°27'8"E. It covers about 75,000 hectares of land. The two rivers Hindon and Kali (West) form the boundaries of the canal command under study. The alluvium in the area under study consists of unconsolidated sediments of clay, sand, silt, kankar and their admixtures. The area experiences a sub-tropical monsoon

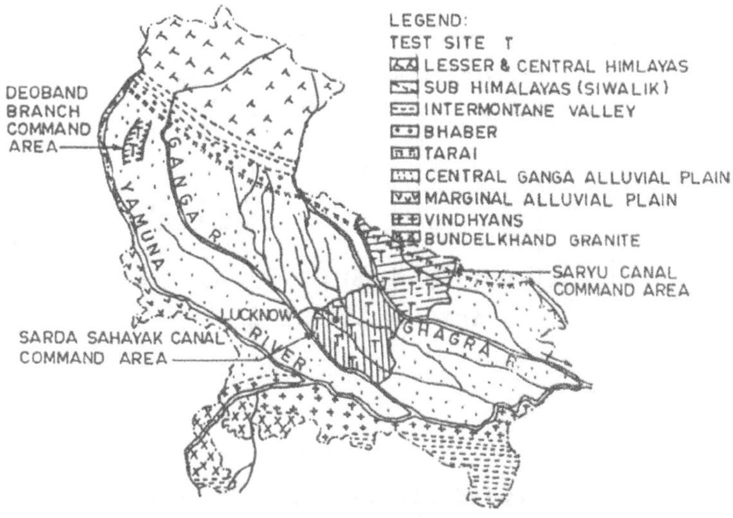

Figure 1 - Location of Test Sites

which is characterised by hot and prolonged summer, followed by rainy season and mild winter. Most of the rainfall occurs during the monsoon months whereas winter rains are scanty. The ground water in these areas has extensive lateral extent and thicknes occur under unconfined or semi-confined conditions.

ISOTOPE TRACER TECHNIQUE

Isotope tracer technique where the water movement is traced directly, tritium in the form of tritiated water (HTO) having a half life of 12.50 years is commonly used as an artificial tracer because the tritiated water behaves like water molecules in the groundwater cycle. The health hazards in handling tritium are also minimum because it emits soft beta particles having maximum energy of 18 Kev only. Moreover the detection sensitivity of the tracer is very high.

The Isotope tracer method of estimating recharge is based on the assumption that soil moisture moves downward in discrete layers. Any fresh layer of water added near the surface, due to applied irrigation, would percolate by pushing an equal amount of water beneath it further down, and so on, such that the moisture of the last layer in the unsaturated zone is added to groundwater. In this technique tritium injected at 70 cm depth will be found as a broad peak perhaps at 80

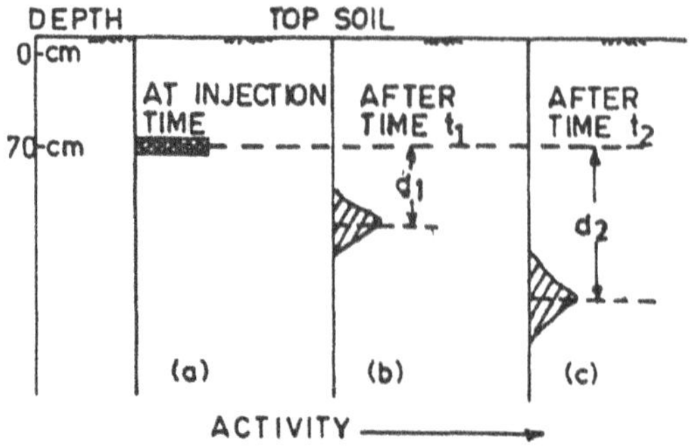

Figure 2. Downward Displacement of Tritium (Expected Profiles)

or 90 cm after a month or so due to some downward movement and some vertical spread due to molecular diffusion and velocity spread. After still longer period, the peak of the activity will move further down and the distribution would be still broader. This is shown diagrammatically in Fig. 2. The position of the tritium peak can be located by counting tritium in the moisture obtained from different depths in soil. Soil samples from different depths can be obtained using hand auger. The moisture content of layer 'd' displaced during the period intervening between the time of injection and time of soil sampling, represents the recharge during the period.

ESTIMATION OF RECHARGE TO GROUNDWATER

The recharge due to applied irrigation in experimental areas was measured by the relationship:

$$Rai = Mvxd \qquad ...(i)$$

Where Rai is the recharge due to applied irrigation, Mv is the average soil moisture content (by vol.) between injection depth and the centre of gravity of the profile and d is the displacement of tracer due to applied irrigation. The rate of moisture movement (Vm) has been determined by the formula;

$$Vm = d/t = \frac{d_2 - d_1}{t_2 - t_1} \qquad ...(ii)$$

Where d is the shift of the tracer peak in time t and d_1 and d_2 are the depths of tracer peak found at time t_1 and t_2 respectively.

FIELD INVESTIGATIONS

Recharge to groundwater due to applied irrigation has been estimated using Isotope tracer technique in three canal commands viz. Deoband branch, Sarda Sahayak canal and Saryu canal under different crops. In Deoband branch command area eight experimental sites viz. Jhabreda,Lakhnota, Deoband, Talheri Buzurg, Muzaffarnagar, Charthawal, Kadargarh and Shahpur were selected for the study. At all the experimental sites, one plot having size 60m x 50m for each crop viz. wheat, paddy and sugarcane were selected as shown in Fig. 1. The selection of these sites were made so as to cover the different hydro-geological characteristics and keeping in view that all facilities of sowing, watering and supervision were available. Moreover these sites were easily approachable. At each experimental site raingauge had been installed for recording the rains during the experimentation.

At all the experimental plots, tritium was injected into the ground at a depth below root zone (70 cm below ground surface) in Dec, 1985 in wheat fields, June, 1985 in paddy fields and April 1985 in sugarcane fields. For injection purposes, small diameter holes were first prepared by driving a pointed rod into the ground upto a depth of 70 cm. At each experimental plot three sets of injection holes at five different places in the plot were made one metre apart and each set consisted of five holes, 10 cm apart to form a square grid. A schematic diagram for injection, implements used and the layout plan in respect of wheat crop for Deoband experimental site is shown in Figs. 3 & 4.

Tritiated water (2 CC) having an activity of 10 μCi were poured into the holes using a piston micro-syringe. After injection the holes were backfilled with the excavated soil. Three sets of injection at each site could ensure collection of more samples at different timings without affecting the natural conditions of soil and moisture movement etc.

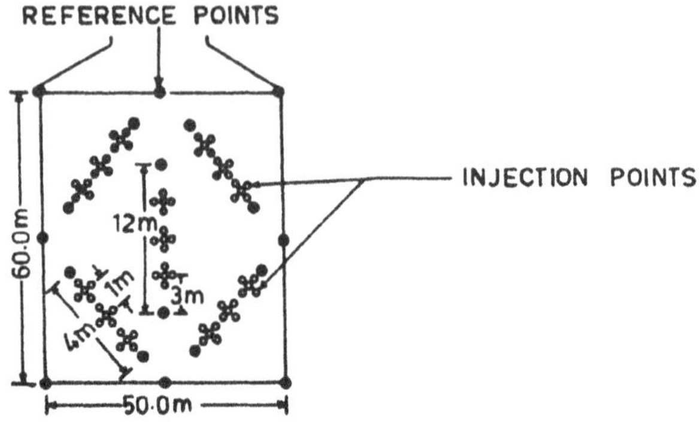

Figure 3—Lay-out Plan of Injection Holes for Wheat Crop at Deoband Experimental Site.

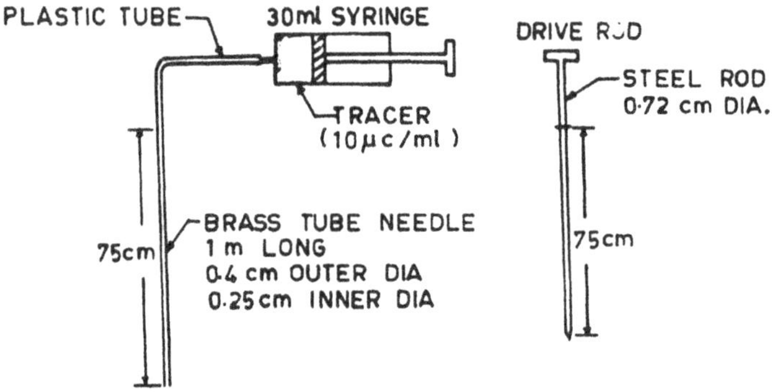

Figure 4— Impliment Used for Injection of Tritium.

Coordinates of each site were noted with respect to trees, electric poles and ruined buildings etc. In addition iron spikes were burried to a depth of about 30 cm and at one metre from all the four corners of injection plan and at all corners of the experimental plot (figure 3). These bearings and marks were used subsequently for locating the injection site for collecting radio-active soil samples. The selected fields were watered, manured and protected with pesticides as per prevailing agricultural practices. The climatological parameters were recorded from district head quarters.

IRRIGATION SCHEDULE AND MODE OF MEASUREMENT

The first irrigation was made after approximately 20 days from sowing in respect of wheat and sugarcane crops while the irrigation was applied before plantation and subsequently in case of paddy crop. The total irrigation applied to each crop is tabulated in Table 1. The soil samples from the above farms at representative locations had been collected after total applied irrigation in successive depths of 15 cm with the help of a hand auger. The sealed plastic containers containing these radio-active soil samples were brought to the laboratory for analysis.

TABLE I - Irrigation Water Applied to Crops by Various Sources

Sl. No.	Test site	Total Amount of Water Applied (Cm)		
		Wheat crop	Paddy crop	Sugarcane crop
1.	Jhabreda	40.40	102.60	122.40
2.	Lakhnota	39.70	98.10	130.60
3.	Deoband	40.10	98.50	130.90
4.	Talheri Buzurg	40.80	97.70	121.80
5.	Muzaffarnagar	45.70	101.90	122.30
6.	Charthawal	42.10	101.40	125.80
7.	Kadargarh	43.60	105.00	126.40
8.	Shahpur	42.40	105.00	125.30
	Average	41.85	101.53	125.70

The quantum of water added to the wheat, paddy and sugarcane fields was measured in the field by constructing the drain from tubewell to experimental fields. The discharge was determined by area-velocity method as well as by volumetric method and total time of irrigation applied was recorded by running a pumping set. Knowing the total time of irrigation applied and discharge of tubewell, the total quantum of water added was determined as below:

$$Ri = Q \times t/A$$

Where Ri is the quantum of water applied in m, Q is the discharge in m^3/sec and t is the total time of irrigation applied in seconds, A is the area of experimental plot in m^2. Insitu bulk density of soil was also determined at each site from the volume and weight of the soil core and given in Table-II.

TABLE II - Average Dry Bulk Density of Soils (gm/cc)

Sl. No.	Test Site	Depth Below Ground Surface		
		0-1 m	1-2 m	2-3 m
1.	Jhabreda	1.60	1.58	1.64
2.	Lakhnota	1.58	1.60	1.60
3.	Deoband	1.65	1.68	1.70
4.	Talheri Buzurg	1.60	1.64	1.65
5.	Muzaffarnagar	1.58	1.60	1.65
6.	Charthawal	1.60	1.64	1.67
7.	Kadargarh	1.65	1.66	1.68
8.	Shahpur	1.64	1.60	1.65

LABORATORY EXPERIMENTS

All the soil samples were again reweighed in the laboratory. In general it was found that there was no loss of soil moisture during transit. The moisture contents of the soil samples were determined by oven dry method at 105 °C. Thereafter each sample was subjected to partial vaccum distillation. One CC of the distillate from each sample was mixed with 15 CC of Dioxane - Nepthalene cock - tail liquid scintillator. The tritium activity of each sample was then counted on an Automatic Liquid Scintillation System (ECIL, Model, LSS-34).

RESULTS AND DISCUSSIONS

The average values of moisture content, Mv (by volume) in between the depth of injection and centre of gravity of the tracer front alongwith those of recharge to groundwater, Rai between sowing and harvesting period from the date of

TABLE III - Recharge to Groundwater Due to Applied Irrigation

Sl. No.	Test Site	Wheat Crop			Paddy crop			Sugarcane crop		
		Moisture (Mv) %	Displacement of tritium (d) cm	Recharge (Rai) cm	Moisture (Mv) %	Displacement of tritium (d) cm	Recharge (Rai) cm	Moisture (Mv) %	Displacement of tritium (d) cm	Recharge (Rai) cm
1.	Jhabreda.	10.24	121.50	12.44	32.26	163.40	52.71	30.75	150.30	46.22
2.	Lakhnota	10.00	118.50	11.85	34.51	159.50	55.04	39.40	112.50	44.33
3.	Deoband	9.54	119.50	11.40	33.33	167.40	55.79	31.20	150.30	46.89
4.	Talheri Buzurg	9.64	121.00	11.66	33.53	170.50	57.17	24.68	160.40	39.59
5.	Muzaffarnagar	11.54	120.00	13.85	34.69	165.30	57.34	29.78	112.00	33.35
6.	Charthawal	10.34	126.00	13.03	35.18	167.20	58.82	38.30	118.80	45.50
7.	Kadargarh	9.28	118.00	10.95	36.49	158.30	57.76	31.00	128.50	39.84
8.	Shahpur	10.80	119.50	12.91	35.40	168.40	59.61	35.40	123.40	43.68
1.	Average recharge		12.26			56.78				42.43
2.	Average total irrigation water applied (Ri) in cm		41.85			101.53				125.70
3.	Average Recharge (Rai) in % (Rai/Ri)		29.30			55.92				33.75

injection have all been summarised in Table III. The tritium activity for soil cores at depth of each 15 cm was plotted against depth for all the eight sites. A typical tritium movement profiles in respect of wheat, paddy and sugarcane crops for Deoband experimental site is shown in Fig. 5.

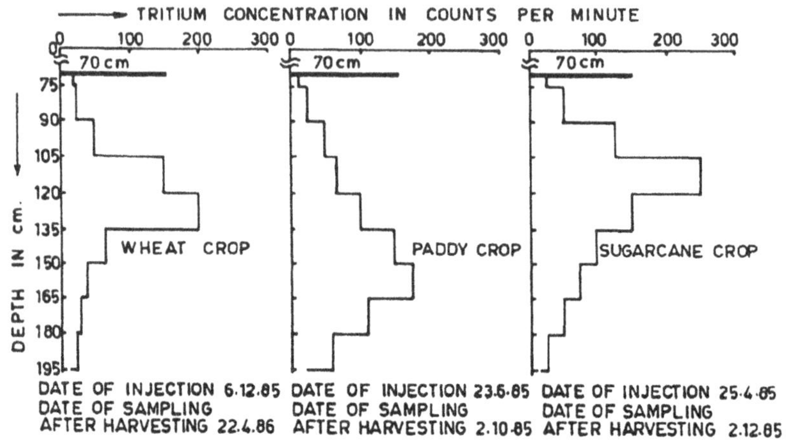

Figure 5: Tritium Movement Profiles in Respect of Wheat, Paddy and Sugarcane at Deoband Experimental Site.

The recharge due to applied irrigation in Deoband branch command area as determined by Isotope tracer technique varies from 25 to 31% for wheat crop, 51 to 59% for paddy crop and 27 to 38% for sugarcane crop respectively. The values for Sarda Sahayak command area and Saryu canal command area are reported in Table IV.

TABLE IV

Sl. No.	Area	Crop	Recharge to groundwater due to applied irrigation (Rai)%
1.	Deoband Branch Canal Command Area	Wheat	25 to 31
		Paddy	51 to 59
		Sugarcane	27 to 38
2.	Sarda Sahayak Canal Command Area	Wheat	27 to 32
		Paddy	52 to 59
		Sugarcane	33 to 38
3.	Saryu Canal Command Area	Wheat	25 to 32
		Paddy	52 to 58
		Sugarcane	34 to 40

Recharge due to applied irrigation in respect of paddy crop has also been estimated by other investigators in various regions and compared fairly well with the results of present study as shown in Table V.

TABLE V - Comparative Studies carried out by various investigators.

Sl. No.	Area of Study	% of Recharge Due to Applied Irrigation	Type of Soil	Method Used
1.	Andhra Pradesh	55 to 88	Sand & Silt Clay	Drum culture method (Lysimeter)
2.	Tamil Nadu	66	Red soil with kankar	-do-
3.	Vattar Karanpudar Canal Command Area Tamil Nadu	65 to 88	-do-	-do-
4.	Deoband branch canal command Area	51 to 59	Silt clay	Isotope Tracer Technique
5.	Sarda Sahayak Command Area	52 to 59	-do-	-do-
6.	Saryu Canal Command Area	52 to 58	-do-	-do-

Radio-active tracer technique provides a fairly acceptable values of recharge due to applied irrigation and are in close agreement with other methods. Moreover, the cost and time involved is relatively less in comparison with conventional techniques.

REFERENCES

1. Agrawal,C.K., Bhargava,A.N., Singh,P.R., Chhabra,S.S. and Suraj Prakash: "Application of Tritium to Study Downward Movement of Soil Moisture in Jamwari-Chauka Doab - A Case Study, Bulletin of Radiation Protection, Vol. 13, No. 1, Jan-March 1990.

2. Datta, P.S., Goel,P.S., Rama and Sangal,S.P.; Groundwater Recharge in Western Uttar Pradesh. Proc. of Indian Acad. Sciences Vol. LXXVIII Sec. A No. 1.

3. Garg,J.M., Devendra Kumar, Anil Kumar and Chhabra,S.S.: Estimation of Groundwater Recharge by Isotopic Method, Proc. of 49th Annual Research and Development Session of C.B.I.& P., 1981.

4. Goel, P.S., Datta,P.S. and Tanwar,B.S.: Measurement of Vertical Recharge to Groundwater in Haryana State (India) Using Tritium Tracer. Nordic Hydrology-8, pp. 221-224, 1977.

5. Jagdish Bahadur et.al; Soil Moisture Movement and Groundwater Recharge by Tritium Tagging Technique, Proc. of Indian Acad. of Sciences, Vol. XXXVA, No.6, pp. 462-471, 1977.

6. Munich,K.D.: Use of Nuclear Technique for the Determination of Groundwater Recharge Rates, Chapter V, Guide Book on Nuclear Techniques in Hydrology, I.A.E.A., Vienna, 1968.

7. Nigam,D.D., Saxena,R.R. and Bhisham Kumar; A method of calculating Recharge to Groundwater from to or Rivers using Tritium", Proceedings of the Workshop on Nuclear Techniques in Hydrology, March, 1980.

8. Zimmermann,U.et.al; Tritium Tagging of Soil Moisture The Water Balance of Forest Soils, I.A.E.A., Vienna.

HYDROGEOLOGICAL CONDITIONS IN THE TERAI PLAIN OF RUPANDEHI DISTRICT, LUMBINI ZONE, NEPAL WITH SPECIAL EMPHASIS ON GROUND-WATER RECHARGE

G. KRISHNA RAO[1], RAMA SHRESTHA[2] and Y.L. VAIDYA[2]

1. Department of Geology, Andhra University, Visakhapatnam - 530 003, India.

2. Groundwater Resources Development Project, Department of Irrigation, Babar Mahal, Kothmandu, Nepal.

ABSTRACT. Terai Plain along the foothill of Himalayas forms a potential zone for groundwater resource. It is generally believed that the confined aquifer system in the Terai Plain is recharged from the Bhabhar zone situated at the foot of the Siwalik (Churia) hills. This aspect is analysed from the fluid potential distribution of the confined aquifer system. The piezometric surface is found to be undulating both in north-south and east-west directions. The rise of pressure head in the southern part in comparison to the central part do not subscribe to the concept of a single recharge source on the north. Palaeochannels in the Terai Plain are found to be acting as recharge zones. The conventional method of groundwater recharge estimate from rainfall in-filtration is not appropriate to the study area. The realistic approach would be the deferred perennial yield method taking into consideration the induced recharge from the present and palaeo-stream beds.

Introduction

The Terai Plain in Rupandehi district, Lumbini zone, Western Region of Nepal has been thoroughly explored for groundwater potential by the U.S. Geological Survey under USAID, The TAHAL group (Groundwater consultants from Israel), Groundwater Resource Development Board, HMG, Nepal and Bhairawa - Lumbini Groundwater Irrigation Project (BLGWP), HMG/N since 1969. The feasibility studies by TAHAL group revealed a groundwater potential of 130 MCM. It is proposed to irrigate 9200 ha with a demand of 8,000 m^2/ha of groundwater through 73 deep tube wells. The basic data and information for this paper has been taken from the Technical Report No. 11 of the GWRDB-UNDP Project prepared by Kunwor and Karanjac (1989).

The Terai Plain in Rupandehi district occupies an area of 1140 km^2 out of which 810 km^2 is considered to be suitable for

V. P. Singh and B. Kumar (eds.), Subsurface-Water Hydrology, 131–149.

irrigation. Groundwater development has been restricted to the contour of 150 m msl which is considered to mark the end of Quaternary sediments in the Terai Plain. The area has the geographical coordinates of 27°48'N Lat. and 83°30'E Long. (Fig. 1). The area enjoys sub-tropical climate with mean monthly temperatures ranging from 15.1°C in January to 30.7° in June. Mean annual rainfall is about 1500 mm which is monsoonal type and 85% of it ensues between June and September. Topographically, the study area is a plain country between 95 and 150 m msl. An upland forms above 150 m msl along the foothills. The plain slopes gently towards south. The gradient is relatively steeper in the northern part at 1:200 while it becomes 1:800 towards southern end. There is an exception to the regional trend in the southeastern part where the ground elevation rises locally to 140 m msl while the surrounding area is at 104 m msl. Tinau river is the major river in the study area which enters the plain from the Himalayas near Butwal and flows across the plain from north to south in the central part.

Hydrogeological Setting

The Terai plain marks the northern fringe of the Gangetic plain. It is divided into Bhabhar zone, middle Terai and southern Terai (Sharma, 1990). But in the study area it may be broadly classified into Bhabhar zone and Terai plain. The Terai plain is underlain by alluvial deposits of Holocene age which also include channel sand and gravel deposits and outwash deposits. The streams in the Terai are characterised by frequent shifting of their courses sometimes by a few kilometers. As a consequence, the sediments are cross-bedded, eroded reworked and redeposited. Aquifer materials show good sorting.

Bhabhar zone occupies relatively a large part in the Rupandehi district with about 130 km^2 along the foot hills of Churia (Siwalik) hills. It represents the river fan and colluvial deposits consisting poorly sorted sediments of boulders, cobbles, gravel and sand. It forms high ground above 150 m msl. The alluvial sediments in the plain are mainly clay and silts intercalated with layers and lenses of gravel and sand. Regionally, the sediments show variation in grain size from north to south becoming finer towards south.

Lithologs of the tube wells drilled to a maximum depth of 300 m with one or two exceptions reveal the subsurface behaviour of the sediment distribution. Three E-W cross-sections of lithologs covering the entire area (Fig. 2) reveal highly irregular and localised nature of sediments. In the northernmost part the sediments vary in size from west to east. In the western part along this section nearby a distributory of Tinau river, the sediments are alternate layers of clays and sand with gravel. In some layers clays are mixed or interlayered with sand. In some layers of gravel, peebles and boulders are also present. Sand is of coarse nature. In the central part close to the Tinau main river, the sediments are gravel, pebbles and boulders with

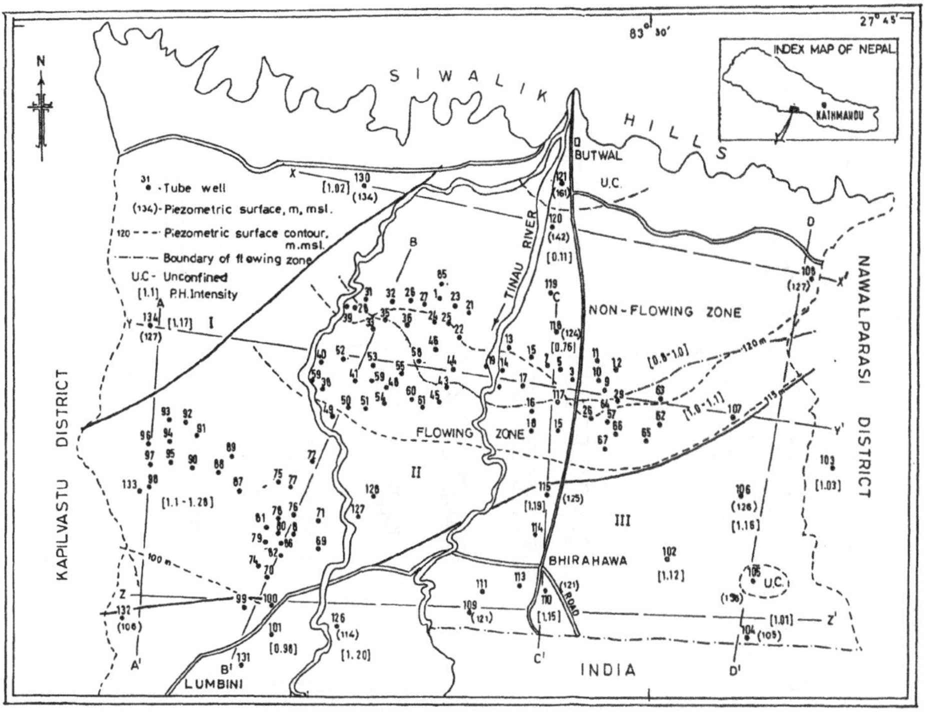

Fig. 1 Location map of Tube wells with some hydrological data, Rupandehi district, Nepal

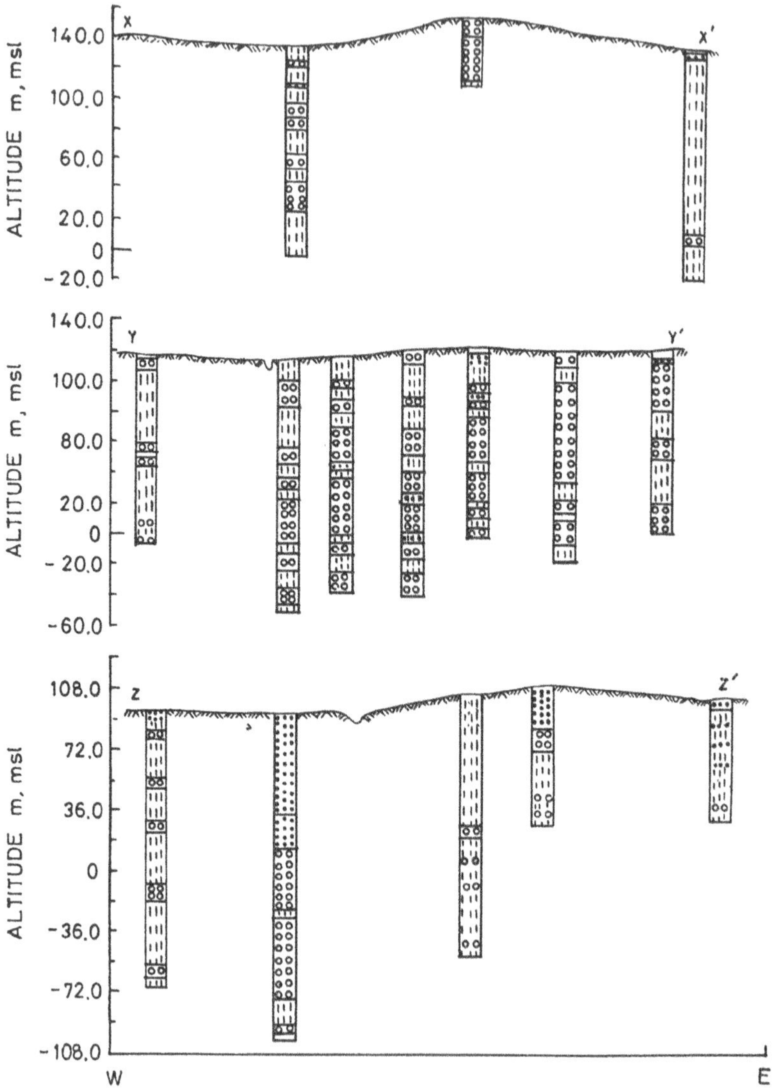

FIG. 2 LITHOLOGICAL CROSS SECTIONS IN E-W DIRECTION, RUPANDEH DISTRICT, NEPAL

(SOURCE : TECHNICAL REPORT No. 11, KUNWOR AND KARANJAC, 1989)

coarse sand and with intermittent thin layers of clay upto a depth of 46 m. In the eastern part which falls in the river basin of Rohini descending from the Churia hills, the sediments are mainly clays with intermittent layers of sand and gravel.

In the central part of the area, the above pattern of east-west variations are observed. In the westermost part, the sediments are predominantly clayey with intermittent thin layers of sand and gravel. In the central part the sediments are predominantly sand and gravel with intermittent thin layers of pure sand and clay or their admixtures. Thickness of clay zones relatively increased on the eastern part. In the southernmost part, the sediment nature distinctly varies from the northern parts. The sediments along this section are predominently clayey with intermittent layers of sand and gravel. Exception is at a place close to the distributory of the Tinau river where it is gravel at depth below 80 m. It is overlain by sand between 60 and 80 m and is mixed with clay towards surface. Interestingly, the sediments are clay rich at depth and sandy on the surface near the main river. On the eastern part, the sediments are sandy.

In the depth range of less than 100 to about 200 m drilled for tube wells, aquifer zones constitute a thickness of 5 to 70 m or slightly more. The normal thickness is 50-60 m. The lithological sections reveal that the aquifers are highly localised in the form of lenses and layers. This is due to frequent shifting of river courses (Sharma, 1974).

AQUIFER CHARACTERISTICS

Free flow from the tube wells touches a maximum of 400 m^3/hr with a static pressure head of about 10 m above ground level. Pumping discharge reaches a maximum of 600 m^3/hr in the project area. However, specific capacity of these wells shows a wide variation from 50 to 4000 lpm/m of drawdown (Table-1) with an average of 1200 lpm/m of d.d. Transmissivity of the aquifer zones is in the range of 100 and 21,000 m^2/day while hydraulic conductivity is 3.5 to 595 m/day (Table-1). In general, the transmissivity falls in the range of 1000 to 15,000 m^2/day. Low values of these parameters mainly confine to the southern part with a few exceptions.

PIEZOMETRIC SURFACE

Most of the area except the Bhabhar zone in the north forms a flowing zone with piezometric heads varying from less than a meter to 16.5 m agl (Table-1). In the central zone the demarcation of flowing and non-flowing wells follows approximately 120 m msl contour line with the flowing wells situated below this contour. In the western part, flowing condition occurs even at 133 m msl. In the non-flowing zone, piezometric surface (P.S) lies at a depth range of less than 1.0 m to 12.5 m.bgl. Unconfined condition is observed in the northern most part near Butwal at an altitude of 174 m msl. Water table lies at 12.8 m bgl.

TABLE I : Hydrological data of tube wells in Rupandehi District, Lumbini zone, Nepal.

S.No	Well No in map	Well No. Field reference	Location	Ground elevation m, msl	Total depth m,bgl	SWL (Piezo-Surface) m,bgl (+ agl)	Piezo-head m	Transmi -ssivity T, m²/day	Hydraulic conductivity K, m/day	Sp.capa lp m/m of d.d
(1)	(2)	(3)	(4)	(5)	(6)	(7)	(8)	(9)	(10)	(11)
1	1	$BL_1/1$	Farsatikar	126.6	115.3	6.6	37.4	16000	355.5	20.3
2	2	$BL_1/2$	Jura (1)	116.6	122.3	+ 4.6	53.6	--	--	2.6
3	3	$BL_1/3$	Bhalwari	121.3	118.6	0.5	44.6	1000	22.7	27.9
4	4	$BL_1/4$	Sakhuwani	119.0	126.0	+ 3.0	51.2	2000	31.2	6.7
5	5	$BL_1/5$	Parshawal	121.7	108.0	+ 1.0	39.7	4000	80.0	12.9
6	6	$BL_1/6$	Kotihawa	116.3	120.0	+ 3.5	52.2	--	--	--
7	7	$BL_1/7$	Karaunjia	121.6	114.4	1.6	36.5	10000	217.4	31.2
8	9	$BL_1/9$	Muriyari	121.7	120.2	1.3	37.7	11500	185.6	57.2
9	10	$BL_1/10$	Jahada	120.7	110.1	0.9	31.3	600	10.5	40.2
10	11	$BL_1/11$	Karaiya	123.1	116.2	1.7	16.0	3000	51.7	29.9
11	12	$BL_1/12$	Semra	120.7	101.4	1.4	30.0	2000	35.7	22.9
12	13	$BL_1/13$	W. Pauni	124.9	110.4	4.4	41.6	12000	24.0	34.6
13	14	$BL_1/14$	Bankathi	123.9	119.0	4.7	40.4	9000	243.2	29.3
14	15	$BL_1/15$	East Pauni	123.8	138.7	2.4	47.9	17000	340.0	8.7

(Source : Technical Report No. 11, Rupendehi district, Deep Ground Water System, Kunwor and Karanjac, 1989).

(1)	(2)	(3)	(4)	(5)	(6)	(7)	(8)	(9)	(10)	(11)
15	16	BL_1/16	N. Supauli	118.2	126.0	+ 0.9	47.9	18500	324.6	14.3
16	17	BL_1/17	Khairiya	121.1	124.0	0.9	44.8	10500	210.0	18.1
17	18	BL_1/18	S. Supauli	115.2	118.5	+ 2.4	50.5	18500	355.0	22.1
18	19	BL_1/19	E. Semra	123.1	114.0	4.8	49.2	6500	144.4	13.8
19	20	BL_1/20	Tikuligarh	121.6	118.0	1.2	51.8	6500	127.5	27.5
20	21	BL_1/21	Dubauliya	127.6	140.0	8.0	45.0	6500	110.2	12.3
21	22	BL_1/22	Bethai	123.5	140.0	3.5	51.5	16500	317.3	16.5
22	23	BL_1/23	Dagna	126.4	121.0	7.7	36.3	5300	101.3	25.5
23	24	BL_1/24	Sude,Farsatikar	122.0	130.0	3.5	41.5	12000	235.3	37.9
24	25	BL_1/25	Mohanganj	122.5	132.0	3.2	43.8	5400	93.1	29.9
25	26	BL_1/26	W. Kewalpur	121.3	142.0	1.4	43.6	18000	352.9	25.4
26	27	BL_1/27	E. Kewalpur	123.9	137.5	4.3	23.7	15900	305.8	15.9
27	28	BL_1/28	Bihuli	117.6	130.2	+ 3.9	48.9	8000	153.8	33.8
28	29	BL_1/29	Pakadihawa	119.0	137.5	+ 3.0	49.0	17700	402.3	37.3
29	30	BL_1/30	N. Gurauliya	115.8	148.7	+ 1.9	43.9			26.9
30	31	BL_1/31	Puraini	117.9	136.0	0.9	40.6	10700	187.7	25.1
31	32	BL_1/32	Rajahar	119.4	136.0	1.80	53.6	15500	271.9	40.5
32	33	BL_1/33	Manpakadi	115.9	143.3	+ 0.62	46.9	8500	141.7	45.1
33	34	BL_1/34	Gurauli-Bhair	112.6	164.0	+ 4.16	60.2	17100	247.8	27.8
34	35	BL_1/35	Sitalpat	117.7	150.6	+ 0.46	47.5	12500	189.4	43.5
35	36	BL_1/36	E. Sitalpat	118.7	184.0	0.24	44.8	12000	171.4	22.6
36	37	BL_1/37	Kunwari	114.6	162.0	+ 3.2	56.2	13800	209.1	41.3

(1)	(2)	(3)	(4)	(5)	(6)	(7)	(8)	(9)	(10)	(11)
37	38	BL_1/38	Madhubani	109.4	191.0	+ 5.95	64.0	9700	127.6	27.6
38	39	BL_1/39	Siswa	110.7	164.0	+ 5.9	49.9	6000	78.9	16.2
39	40	BL_1/40	Belahia	112.1	176.0	+ 5.6	55.6	7000	120.7	4.6
40	41	BL_1/41	Bhaglapur	110.3	176.0	+ 6.2	49.2	7100	101.4	45.5
41	42	BL_1/42	Kotihawa	116.6	168.0	+ 3.5	53.5	16700	228.8	38.5
42	43	BL_1/43	Amua	117.4	164.0	+ 0.9	47.9	13900	190.4	22.4
43	44	BL_1/44	W. Sekhuwani	118.9	161.0	0.65	50.3	11600	175.8	57.5
44	45	BL_1/45	Kanari	115.5	164.0	+ 1.9	57.9	7700	118.5	8.9
45	46	BL_1/46	Sarjugunj	123.9	149.3	5.14	50.9	6000	98.4	35.6
46	47	BL_1/47	Rehara	120.3	160.0	1.7	45.3	6100	91.0	65.7
47	48	BL_1/48	Ramganj	114.9	163.0	+ 0.1	53.1	5700	103.6	27.7
48	49	BL_1/49	W.Khungaon	109.8	168.0	+ 3.0	55.0	3400	57.6	13.0
49	50	BL_1/50	E.Khungaon	111.9	162.0	+ 3.14	53.1	2000	29.4	10.4
50	51	BL_1/51	Dayanagar	112.6	164.0	+ 2.49	56.5	5200	76.5	17.8
51	52	BL_1/52	W. Semari	113.3	159.0	+ 3.2	54.2	17500	246.5	30.8
52	53	BL_1/53	Bahata	114.9	155.0	+ 2.7	48.7	2500	34.7	4.4
53	54	BL_1/55	Chapia	113.7	140.0	+ 0.98	63.0	5700	100.0	25.4
54	55	BL_1/56	Madagunj	117.2	160.0	0.48	38.5	7600	135.7	21.1
55	56	BL_1/57	Juda	116.6	153.0	+ 2.95	47.0	3900	66.1	15.4
56	57	BL_1/58	Bharauli	118.7	160.0	0.04	44.9	11100	213.5	35.9
57	58	BL_1/59	S.Madagunj	114.7	158.0	+ 2.8	47.8	21000	403.8	39.5
58	59	BL_1/60	Banghosary	112.9	160.0	+ 3.3	50.3	9000	152.5	21.2
59	60	BL_1/61	Baidauli	113.9	162.0	+ 2.03	50.1	14100	243.1	40.7

(1)	(2)	(3)	(4)	(5)	(6)	(7)	(8)	(9)	(10)	(11)
60	61	BL_1/62	Bardahawa	113.9	163.0	+ 4.14	64.1	2400	40.7	10.4
61	62	BL_1/63	Jamuhani	116.9	161.0	+ 3.87	73.9	3200	50.0	5.7
62	63	BL_1/64	Beuratola	117.6	158.0	+ 1.99	49.0	14200	284.0	43.0
63	64	BL_1/65	Chamkipur	118.5	157.0	+ 2.6	52.6	13800	194.4	30.6
64	65	BL_1/66	Tataria	116.2	172.0	+ 1.8	63.8	10000	156.4	11.3
65	66	BL_1/67	Barauliya	116.9	169.0	+ 2.4	52.4	14000	241.4	21.9
66	67	BL_1/68	Gongolia	115.9	158.0	+ 1.6	51.6	13600	234.5	25.0
67	68	BL_1/69	Patharda	118.9	150.0	+ 0.07	61.1	18600	315.5	51.6
68	69	BL_{11}/1	Harharwa	95.9	193.0	+ 6.8	70.8	14091	242.9	6.0
69	70	BL_{11}/2	Pipra	92.8	194.0	+ 8.86	68.9	16426	283.2	3.2
70	71	BL_{11}/3	Bhilarahwa	97.3	163.1	+ 6.58	68.6	14091	299.8	3.8
71	72	BL_{11}/4	Mahadava	99.3	203.0	+ 8.48	51.5	1780	32.9	6.4
72	74	BL_{11}/5	Ekla	94.8	180.1	+ 6.6	63.6	6718	101.8	6.0
73	75	BL_{11}/6	Manglapur	98.9	182.0	+ 3.7	52.7	2050	50.0	2.2
74	76	BL_{11}/7	N. Ahirauli	100.7	163.0	+ 7.66	53.7	11359	206.5	6.6
75	77	BL_{11}/8	Amauli	98.7	167.0	+ 7.14	54.1	2221	46.3	6.4
76	78	BL_{11}/9	Gobrauli	98.1	166.1	+ 9.5	38.4	2636	54.9	5.9
77	79	BL_{11}/10	Bhagatpur	95.9	195.0	+ 4.55	69.6	6718	129.2	2.4
78	80	BL_{11}/11	E.Bharauliya	96.9	222.0	+ 7.22	99.2	8622	172.4	1.9
79	81	BL_{11}/12	W.Bharaulia	98.5	198.0	+ 6.56	87.6	8622	233.0	1.6
80	82	BL_{11}/13	Sonbarsa	94.2	206.0	+ 5.98	63.0	1412	23.2	5.6
81	84	BL_{11}/14	S. Ahirauli	96.1	171.0	+ 7.40	51.4	5113	86.7	7.4

(1)	(2)	(3)	(4)	(5)	(6)	(7)	(8)	(9)	(10)	(11)
82	85	BL_{11}/15	Mainaiya	--	156.0	10.48	33.5	6154	94.7	29.9
83	86	BL_{11}/16	S.Bharaulia	96.3	201.0	+ 4.54	53.5	3331	64.1	2.1
84	88	BL_{11}/18	Kurmindihawa	100.1	164.0	+ 5.80	50.9	367	7.9	7.6
85	89	BL_{11}/19	Dhauthaniya	100.1	163.0	+ 8.60	33.6	334	6.4	0.8
86	90	BL_{11}/20	Gidahiya	98.1	150.0	+ 8.84	67.8	3938	75.7	1.1
87	91	BL_{11}/21	Pauwa	99.4	175.0	+ 8.01	60.0	1123	26.7	0.1
88	92	BL_{11}/22	N. Pauwa	104.2	152.0	+ 1.70	54.7	1883	52.3	10.5
89	96	BL_{11}/26	S. Sihawa	102.2	163.0	+ 2.48	48.5	831	20.3	15.8
90	101	BL_1NV/1	Patijiya	--	213.5	1.22	51.8	250	8.3	0.6
91	102	BL_1NV/3	Dhakdahi	--	185.0	+ 3.27	31.3	2471	61.8	2.2
92	103	BL_1NV/2	Banjaria	--	187.0	+ 2.80	108.8	253	6.3	4.2
93	104	U34/1	Vishnupura	104.0	72.8	+ 1.00	64.0	160	22.9	25.0
94	106	U34/4	Belahia	114.0	79.3	+ 12.0	88.0	851	20.9	--
95	108	U34/6	Kerwani	135.0	152.0	8.0	114.0	400	26.7	30.3
96	109	U35/3	Paklihawa	104.2	156.0	+ 16.5	163.5	105	3.5	2.0
97	110	U35/4	Bhairahawa (S. P.Camp)	110.0	82.0	+ 10.67	83.7	692	69.2	10.8
98	115	U35/10	Govt.Fish Farm	115.8	55.0	+ 9.2	57.2	2396	399.3	54.0
99	118	U35/14	Manigram	128.6	48.0	3.35	10.6	6661	595	41.2
100	120	U35/18	Jogikuti	154.2	46.0	12.50	1.5	2805	561	20.8
101	121	U35/19	Butwal	174.0	90.0	12.80	-0.00	1412	94.1	6.5
102	126	U56/7	Semri	100.6	461.0	+ 13.1	78.1	878	176	16.7

(1)	(2)	(3)	(4)	(5)	(6)	(7)	(8)	(9)	(10)	(11)
103	129	U56/12	Bhujauli	120.9	99.0	+ 4.11	49.1	800	80.0	16.7
104	130	U56/13	Chetri	132.6	137.0	+ 1.2	71.2	--	--	--
105	132	U58/3	Mugh	94.5	164.0	+ 11.9	163.9	227	9.1	4.4
106	133	U58/4	Sarahawa	100.5	152.0	+ 3.35	37.4	--	--	--
107	134	U58/5	Asnia	115.8	123.0	+ 11.28	77.3	235	11.8	7.0

Piezometric surface ranges in elevation from 100 to 142 m.msl increasing twoards north and following topography (Fig. 1). On the basis of its configuration, the study area is divided into three zones (I, II and III) from northwest to southeast. The central zone which is a broad one occupies most of the Tinau river basin and lies at a lower elevation than the other two. While the northwestern zone lies outside this basin, the southeastern zone falls in its lower reaches or partly under Rohini river (not shown in the map). Hydraulic gradient in the three zones is from northeast to southwest.

Profiles of piezometric surface in north-south direction (Fig. 3) clearly show its undulating nature with a depression in the central part. The rise in piezometric surface along the southern part compared to the central part is 1.5 to 5.0 m. The piezometric surface on the eastern part (profiles C-C' and D-D') is at higher elevation than the western part.

PIEZOMETRIC HEAD

Piezometric head (P.H) i.e. rise of water level in the tube wells from the top of the upper most aquifer tapped, ranges from 1.5 to 164 m depending on the depth of the aquifer. According to hydraulic principle, P = wh (P is pressure head, w is specific weight of water and h, depth of the aquifer), When w for water is constant, pressure head is directly related to depth of the aquifer. Such direct relationship has been observed in some parts of the Terai plain like Morang district (Batarai and Krishna Rao, 1989), Nawalparasi district (Pathak, 1991) and in the Kathmandu valley (Ramesh Gautam and Krishna Rao, 1991) in case of non-flowing wells. P.H. in case of flowing wells is higher than the above relation. In the study area, the P.H. in most of the wells which are flowing is higher than the above relation (Fig. 4) and it is less in a few cases which are located towards northern part located in semiconfined aquifiers. Intensity of piezometric head i.e. development of head per meter depth of the aquifer is found to range from 0.11 to 1.32 (Fig. 5). Flowing wells result where the intensity is more than 1.0. In the central (zone II), zone, the intensity is observed to increase gradually from north to south. In the northwestern zone (zone I), the intensity is developed to 1.02 even at higher elevation and it further increased to 1.17 towards southwest. The sotheastern zone (zone III) also shows a different trend from the central zone. The intensity is much higher in the southern and southwestern parts. This is evident from the rise of P.S. in these parts. It infers that flowing conditions in the study area is not due to topographic control alone but also due to higher hydraulic head. In some places, the nature of confining materials seems to have some influence on the piezometric head as in the case of wells 5 and 7. Though they are of same depth and at same elevation, the intensity is less where the confining layer is predominently sandy clay in comparison to the other one which is confined with a thick clay bed. In the former case there is a possibility of the pressure dissipation through the overlying semipermeable beds.

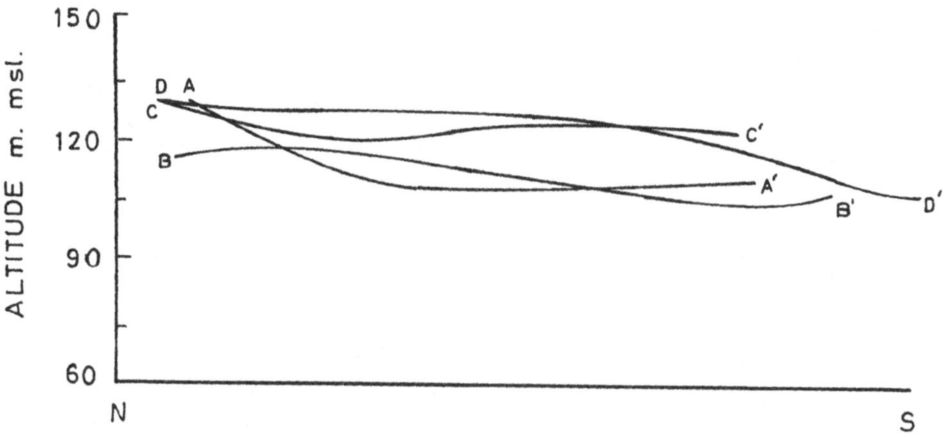

FIG. 3 PIEZOMETRIC SURFACE PROFILES IN N-S DIRECTION RUPANDEHI DISTRICT, NEPAL.

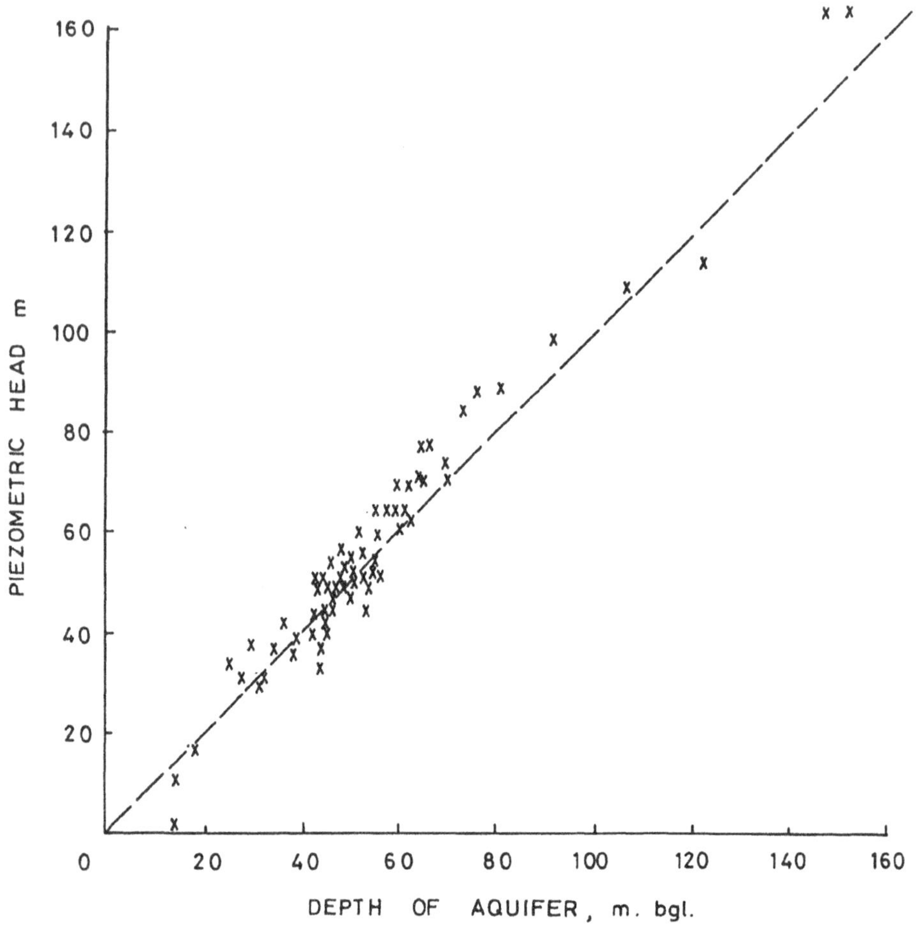

FIG. 4 RELATION BETWEEN PIEZOMETRIC HEAD AND DEPTH OF AQUIFER, RUPANDEHI DISTRICT, NEPAL

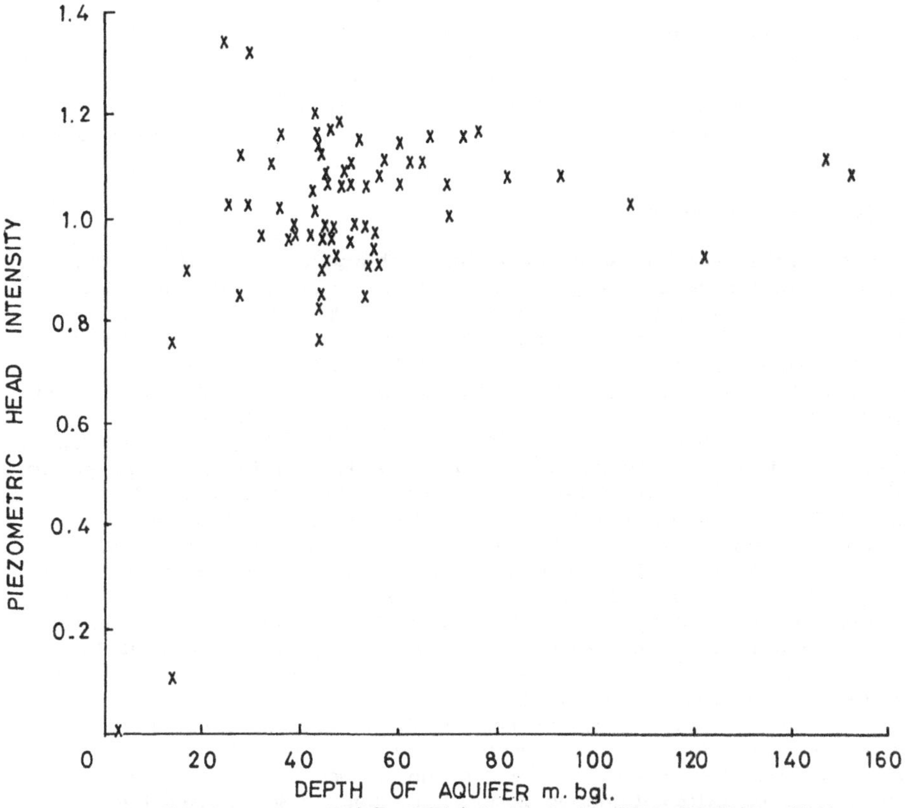

FIG. 5 TUBE WELLS IN RUPANDEHI DISTRICT, NEPAL

RELATION BETWEEN PIEZOMETRIC HEAD INTENSITY AND DEPTH OF AQUIFER

A graph plot of piezometric head intensity versus depth of aquifer (Fig. 5) shows lack of any significant correlation in the study area. Maximum intensity is obtained between 25 and 75 m and it is not increased further even down to 150 m. Nonflowing condition is found to exist in the aquifer even at a depth of 122 m in some places while flowing conditions are encountered beyond 25 m depth at many places.

Groundwater Recharge

Bhabhar zone is considered to be the recharge area for the confined aquifers in the Terai plain (Karanth, 1989; Kunwor and Karanjac, 1989). Duba (1982, in Kunwor and Karanjac, **Op. cit**) estimated that 43.7% of rainfall in the Bhabhar zone goes as groundwater recharge which comes to 119 MCM (Million Cubic Meters). A conservative estimate with 25% of rainfall as infiltration rate into the deep aquifer system puts it at 68 MCM (Kunwor and Karanjac, **Op.cit**).

The undulating nature of piezometric surface from north to south, however, does not support the above concept. Applying the principle of friction of flow, the piezometric head is expected to decline gradually from north to south. A rought estimate of friction losses through gravelly sediments shows it as 0.1 m/m length of flow. The fact that the piezometric surface rises in the southern part (zone III) in comparison to central part (zone II) cannot be explained by this single source concept. Presence of sand, gravel and boulder sediments in the southeastern part (Well No. 105) forming high ground at 140 m msl in the midst of fine grained sediments represent a palaeo-stream course. This area, therefore, is acting as another recharge zone in the southern part. Similarly, the northwestern part (zone I) which has a different configuration of piezometric surface distinct from the central zone may be considered to be under the influence of a different recharge area either Bhabhar zone or another palaeo-stream bed still at higher altitude at the foot hills of the Siwaliks. A similar situation is observed in the adjoining area on the eastern part in Nawalparasi district. In this area, the piezometric surface along the Turia river course is relatively at higher elevation than the remaining part of the district and inferred to have its recharge area at the foothills of Siwaliks along the river course (Krishna Rao and Pathak, 1993). The higher elevation of piezometric surface on the eastern part of the study area possibly forms a continguous part of this zone and hence may be considered to be under the recharge influence of Rohini river situated between Tinau and Turia rivers.

ANNUAL RECHARGE

According to a study in the Bhairahawa-Lumbini Groundwater Project (Kunwar and Karanjac, **Op. cit**), the total discharge of groundwater during 1983-84 through pumps and free flow was 56.7 MCM and it increased to 59.9 MCM during 1985-86. The effect of this increase in draft is observed on the water levels of the deep tube wells when

TABLE II : Seasonal water level fluctuations in tube wells during 1983 and 1988, Rupandehi district, Lumbini zone, Nepal

S.No	Well No. & Location	Altitude (m, msl)	Water level, m, msl						Difference in June 1983 to 1988 (m)	Difference in October 1983 to 1988 (m)
			June, 1983	October, 1983	Rise m	June, 1988	October, 1988	Rise m		
1	121 Butwal	174	158	168	10	159.0	168.5	9.5	+ 1	+ 0.5
2	1 Farsatikar	126.6	118.2	119.3	1.1	115.7	118.5	2.8	--2.5	--0.8
3	13 W. Pauni	125.0	119.5	120.8	1.3	116.5	120.0	3.5	- 3.0	- 0.8
4	20 Tikuligarh	121.6	118	119.0	1.0	116.5	117.5	1.0	- 1.5	- 1.5
5	53 Bahata	114.9	117.5	119.0	1.5	113.5	116.5	3.0	- 4.0	- 2.5
6	65 Tataria	116.2	120.4	120.8	0.4	118.5	120.5	2.0	- 1.9	- 0.3

compared for 1983 and 1988 (Table II). Except at Butwal which is located in the Bhabhar zone, all the tube wells in the Teria plain show a decline of pre-monsoon (June) water levels from 1983 to 1988 by 1.5 to 4.0 m. The post-monsoon (October) levels also record a fall of 0.3 to 2.5 m. But it is interesting to note that when the annual fluctuation (rise from June to October) is 0.4 to 1.5 m during 1983, it increased to1.0 to 3.5 m during 1988. This shows doubling of the annual recharge in 1988. It shows that the groundwater recharge increased with increase of groundwater draft. This confirms the possibility of induced recharge either through leakage from shallow aquifers or from the present and palaeo-stream beds. Since most of the southern part is a flowing zone with the piezometric surface falling above the ground level, vertical downward infiltration in this belt is not possible. However, infiltration from stream beds may take place tangentially downslope side following hydraulic gradient.

Conclusions

The potentiometric surface of the confined aquifers in the Terai Plain is undulating in north-south and east-west directions. It reveals presence of different recharge areas in different parts in addition to the Bhabhar zone. The present and palaeo-river courses which are abundant in the Terai Plain are active recharge zones.

It is therefore, suggested that the annual groundwater recharge in the Terai Plain may be worked out by subjecting to heavy exploitation initially to lower the piezometric surface and to allow maximum additional induced recharge instead of restricting it to the conventional rainfall recharge estimates.

References

Bhattarai, T.N., Krishna Rao, G., 1989. "Hydrogeological conditions in the Terai plain of Morang District, Eastern Nepal'. **Journ. Nepal, Geol. So.**, Vol. **6**, pp. 37-50.

Gautam, R., Krishna Rao, G., 1991. 'Groundwater Resource evaluation of the Kathmandu Valley'. **Jour. Nepal Geol. Soc.**, Vol. **7**, pp. 39-48.

Karanth, K.R., 1989. **Hydrogeology.** Tata McGraw Hill Publishing Co. Ltd., New Delhi, p. 458.

Krishna Rao, G., Pathak, D., 1993. 'Hydrogeological Conditions in the Terai Plain of Nawalparasi district, Lumbini zone, Nepal, with special reference to groundwater recharge'. Abstract. **Int. Sem. on Hydrology**, held at Kathmandu, Nepal.

Kunwor, M.B., Karanjac, J., 1989. 'Deep tubewells Drilling, Testing and Monitoring in 1969-89, Basic Documentation and Preliminary Interpretation, Rupandehi District, Technical Report No. 11, GWRDB - UNDP Project NEP/86/025.

Pathak, D., 1991. 'Hydrogeological studies in parts of Nawalparasi District, Western Nepal'. M.Sc. Dissertation thesis submitted to the Central Department of Geology, T.U., Kathmandu.

Sharma, C.K. 1974. **Groundwater Resources of Nepal**, p. 40.

Sharma, C.K. 1990. **Geology of Nepal and Adjacent Countries.**

SECTION 4:

AQUIFER RESPONSE

THE ENERGY BALANCE OF GROUNDWATER FLOW.

R.J. Oosterbaan[1], J. Boonstra[1] and K.V.G.K. Rao[2]

[1] International Institute for Land Reclamation and Improvement (ILRI), P.O. Box 45, 6700 Wageningen, The Netherlands.
[2] Central Soil Salinity Research Institute (CSSRI), Karnal-132001, Haryana, India

ABSTRACT. An energy balance of groundwater flow is introduced. It is based on equating the change of hydraulic energy flux over a horizontal distance to the conversion rate of hydraulic energy into to friction of flow over that distance. The energy flux is calculated on the basis of a multiplication of the hydraulic potential with the flow velocity, and is integrated over the total flow depth. The conversion rate is determined in analogy to the heat loss equation of an electric current. The hydraulic energy balance is applied to the steady-state flow of water in a phreatic aquifer recharged by downward percolation stemming from rainfall or irrigation, and a quantitative example is given using a numerical solution. It is shown that the gradient of the watertable is smaller than that calculated with the current methods, which do not take into account the energy associated with the incoming percolation water.

1. ENERGY BALANCES

1.1 Energy fluxes

The hydraulic potential (P) can be defined as the energy per unit volume of water (ϵ/m^3, where ϵ represents energy units). The flow velocity (V) of groundwater can be defined as the discharge per unit cross-sectional area perpendicular to the direction of flow (m^3/day per m^2). The product P.V therefore represents an energy flux, i.e. an energy flow per unit cross-sectional area (ϵ/day per m^2).

Figure 1 shows a longitudinal section of two-dimensional groundwater flow (i.e. the flow pattern repeats itself in the planes parallel to the plane of the drawing) in a phreatic aquifer, i.e. an aquifer with a free watertable. The watertable is recharged by downward percolating water (R m/day) stemming from rainfall and/or irrigation. A coordinate system, with X (m) giving the horizontal and Z (m) the vertical distance from the origin, is also indicated. The horizontal component of the flow velocity in any point (X,Z) is indicated by Vx. The Z-levels of the impermeable layer and the watertable in a vertical cross-section are shown as I and J respectively.

The total energy flow through a vertical cross-section (Ex, ϵ/day per m width in the direction perpendicular to the longitudinal section) is

$$Ex = \int_I^J [Vx(P-Pr)]dZ \qquad (1.1)$$

where Pr is a reference value of P, independent of X and Z, to be determined in accordance to the boundary conditions of the flow.

V. P. Singh and B. Kumar (eds.), Subsurface-Water Hydrology, 153–160.

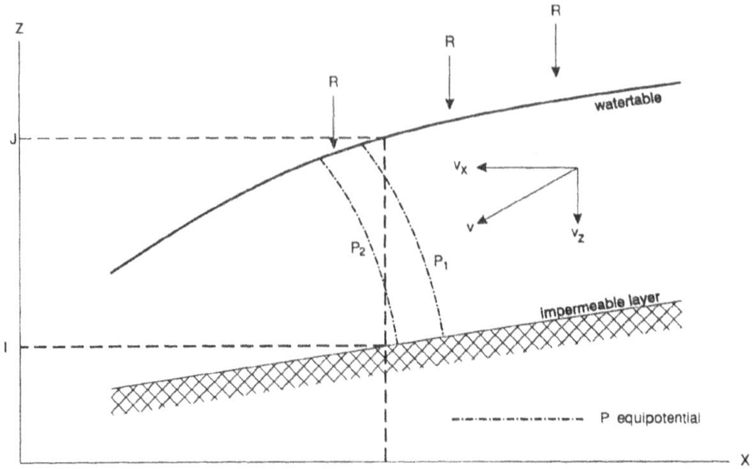

Figure 1. A vertical cross-section in a longitudinal section along two-dimensional groundwater flow in a phreatic aquifer recharged by percolation.

The change of the energy flow Ex per unit distance in a horizontal direction is

$$\frac{dEx}{dX} = \frac{d}{dX}\int_I^J [Vx(P-Pr)]dZ$$

Using Leibnitz's rule, and assuming that the impermeable layer is horizontal (I=0, dI/dX=0), the above equation can be written as

$$\frac{dEx}{dX} = \int_0^J [\frac{d}{dX}(Vx(P-Pr))]dZ + Vj(Pj-Pr)\frac{dJ}{dX}$$

where Vj and Pj are the values of Vx and P at the watertable.

Partial differentiation of the product Vx(P-Pr) in the previous Equation, and noting that dPr/dX=0, yields

$$\frac{dEx}{dX} = \int_0^J (Vx\frac{dP}{dX})dZ + \int_0^J [(P-Pr)\frac{dVx}{dX}]dZ + Vj(Pj-Pr)\frac{dJ}{dX} \qquad (1.2)$$

1.2. The hydraulic head

The energy units ϵ, expressed in S.I. units, are kg.m^2/day^2, so that the units ϵ/m^3 of the potential P become kg/day^2 per m. The potential P can be converted into an hydraulic head as follows

$$H = P/\rho.g$$

where

H is the hydraulic head (m)
ρ is the mass density of water (kg/m^3)
g is the gravitational acceleration (m/day^2)

With the conversion of potential P into head H, Equation 1.2 becomes

$$\frac{dEx/dX}{\rho.g} = \int_0^J (Vx\frac{dH}{dX})dZ + \int_0^J [(H-Hr)\frac{dVx}{dX}]dZ + Vj(Hj-Hr)\frac{dJ}{dX} \qquad (1.3)$$

F4rom elementary hydraulics we know that the head H consists of three components: the elevation head (H_Z=Z), the pressure head (H_Y) and the velocity head (H_V). The velocity head of groundwater flow is negligibly small, so that H=Z+H_Y. At a phreatic surface, i.e. at the free water-table, the pressure head corresponds to atmospheric pressure, which can be taken as zero reference pressure, so that H_Y=0. Hence, for Z=J, i.e. at the watertable, we find $H_{Z=J}$=J.

Using the Dupuit assumptions that the velocity Vx, the head H, and the gradient dVx/dX are constant with height Z, so that Vj=Vx and H=$H_{Z=J}$=J, and writing Jr for Hr, Equation 1.3 can be simplified to

$$\frac{dEx/dX}{\rho.g} = (Vx.J)\frac{dJ}{dX} + J(J-Jr)\frac{dVx}{dX} + Vx(J-Jr)\frac{dJ}{dX} \qquad (1.4)$$

The assumption that the horizontal velocity Vx is constant with height is realistic when the resistance to vertical flow is small compared to that to horizontal flow. The assumption does not imply that the real velocity (V), i.e. the vector sum of horizontal (Vx) and vertical velocity (Vz), is also constant, as Vz need not be constant with Z.

1.3. The water balance

When the velocity Vx is constant with height Z, the two-dimensional discharge Q (m^3/day per m width of cross-section, or m^2/day) equals Q=Vx.J, and its differential coefficient, i.e. the change of discharge Q per unit change in distance X, becomes:

$$\frac{dQ}{dX} = \frac{d(Vx.J)}{dX} = Vx\ \frac{dJ}{dX} + J\ \frac{dVx}{dX} = R$$

Hence, Equation 1.4 can be simplified to:

$$\frac{dEx/dX}{\rho.g} = (Vx.J)\frac{dJ}{dX} + R(J-Jr) \qquad (1.5)$$

1.4. Energy conversion by friction of flow

The electric current in a conduit is known to lose electrical energy by its conversion to heat. The conversion rate is proportional to the resistance of the conduit and the square value of the current. The resistance is inversely proportional to the conductance. In analogy, the conversion of hydraulic energy to friction of flow is taken as

$$Fx = \int_0^J [\frac{(Vx)^2}{Kx}]dZ$$

where Kx is the horizontal hydraulic conductivity of the soil (m/day). Further, it is stated that the energy loss rate (as in Equation 1.5) is proportional to the negative value of the friction losses. Thus we obtain:

$$\frac{dE/dX}{\rho . g} = - Fx$$

Combining the previous two equations, and assuming again that the velocity Vx is constant with height Z, one obtains

$$\frac{dEx/dX}{\rho . g} = - J \frac{(Vx)^2}{Kx} \tag{1.6}$$

1.5. The hydraulic energy balance for steady state flow

When there is no change in storage of water, and consequently there is no change in storage of hydraulic energy (i.e. energy storage associated with water storage), we have a steady state: the hydraulic energy losses are fully converted into frictional energy. It can then be found from Equation 1.5 and 1.6 that

$$Vx \frac{dJ}{dX} + \frac{R(J-Jr)}{J} = - \frac{(Vx)^2}{Kx}$$

The minus sign in the above equation assures that the energy losses are positive when the gradient dJ/dX is negative, which occurs when the flow Vx is positive (i.e. in the positive x-direction or, in Figure 1, to the right), and vice versa. Division by Vx and rearrangement gives:

$$\frac{dJ}{dX} = - \frac{Vx}{Kx} - \frac{R(J-Jr)}{Vx.J} \tag{1.7}$$

2. PHREATIC AQUIFERS WITH RECHARGE

2.1. The hydraulic energy balance equation

Figure 2 shows the two-dimensional flow of groundwater in a phreatic aquifer recharged by evenly distributed percolation from rainfall or irrigation (R>0, m/day). At the distance X=N (m) there is a water divide, here the watertable is horizontal. The impermeable base is taken horizontal. The height of the watertable above the impermeable base is taken equal to J (m). At the distance $X \leq N$, the discharge of the aquifer equals Q=-R(N-X) (m^2/day). We find:

$$Vx = Q/J = -R(N-X)/J$$

With this, Equation 1.7 can be changed into

$$\frac{dJ}{dX} = \frac{R(N-X)}{Kx.J} - \frac{Jr-J}{N-X}$$

Setting F=J-Jo and Fr=Jr-J, where Jo is the value of J at X=0, and applying the condition that dF/dX=0 when X=N, we find Fr=Fn, where Fn is the value of F at X=N, and

$$\frac{dF}{dX} = \frac{R(N-X)}{Kx.J} - \frac{Fn-F}{N-X} \tag{2.1}$$

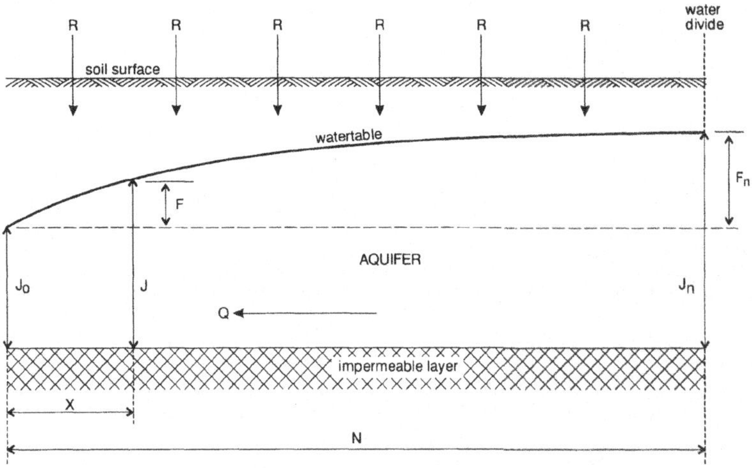

Figure 2. Flow conditions of groundwater in a phreatic aquifer recharged by percolation.

2.2. Integrations

Integrating Equation 2.1 from X=0, where F=0, to any value X, gives

$$F = \int_0^X \left[\frac{R(N-X)}{Kx.J}\right] dX - \int_0^X \left[\frac{Fn-F}{N-X}\right] dX \tag{2.2}$$

The integration of the last term requires advance knowledge of Fn. To overcome this problem, a numerical solution and a trial and error procedure is given in Section 2.4.

2.3. The current method of analysis

When, according to the current method of analysis, the Darcy equation is used with the water balance and the Dupuit assumptions to describe the groundwater flow under the same conditions, one finds instead of Equation 2.2 (e.g. Wesseling 1973):

$$F^* = \int_0^X \left[\frac{R(N-X)}{Kx.J}\right] dX \tag{2.3}$$

Here, the symbol F* is used instead of F to indicate the current method of analysis. In the following, a numerical solution of Equation 2.4 is given, but the equation can also be solved directly as

$$F^* = \tfrac{1}{2}R.N^2/Kx.J \tag{2.4}$$

2.4. Numerical integrations

For the numerical integrations, the horizontal distance N is divided into a number (T) of equal small elements with length U, so that

U=N/T. The elements are numbered S = 1, 2, 3, ..., T. The heights F and J of the watertable in the point defined by the largest value of distance X in element S are denoted as F_S and J_S. The change of height F over the S-th element in the zone of radial flow is denoted as G_S and found from

$$G_S = F_S - F_{S-1}$$

The average value of height F over the S-th element is

$$\underline{F}_S = F_{S-1} + \tfrac{1}{2}G_{S-1}$$

and the average of the cross-sectional height J of flow is

$$\underline{J}_S = J_{S-1} + \tfrac{1}{2}G_{S-1}$$

The average value of the horizontal distance X of the S-th element from the center of the drain is found as

$$X_S = U(S - 0.5)$$

Equation 2.1 can now be approximated by:

$$G_S = U(A_S+B_S) \qquad (2.5)$$

where

$$A_S = R.(N-X_S)/K.\underline{J}_S$$

$$B_S = (\underline{F}_S-F_T)/(N-X_S)$$

where F_T is the value of F_S when S=T. Now, the height of the watertable at any distance X can be found, conform to Equation 2.2, from:

$$F_S = \sum_{1}^{S} G_S \qquad (2.6)$$

Since F_S depends on B_S and B_S on F_T, which is not known in advance, Equation 2.6 must be solved by trial and error.

In similarity to the above procedure, the value G_S* (where the symbol * is used to indicate the numerical solution of Equation 2.3 instead of 2.2, i.e. not using the energy balance but the current method of analysis) is found as

$$G_S{*} = R.U(N-X_S)/K.\underline{J}_S{*} \qquad (2.7)$$

where $\underline{J}_S{*}=J_{S-1}{*}+\tfrac{1}{2}G_{S-1}{*}$. Thus the height of the watertable, in conformity to Equation 2.4, is:

$$F_S{*} = \sum_{1}^{S} G_S{*} \qquad (2.8)$$

2.5. Example of a numerical solution

To illustrate the numerical solutions we use the following data:

N	=	100.0	m	J_0	=	10.0	m
Kx	=	1.0	m/day	R	=	0.01	m/day
U	=	0.5	m				

For the example, the calculations with Equations 2.5, 2.6, 2.7 and 2.8 were made on a computer. The results are presented in the Tables 1 and 2 and in Figure 3.

Table 1 gives the values of height F_S (m) and gradients G_S/p, A_S, B_S at some selected values of distance X, using Equations 2.5 and 2.6 (i.e. using the energy balance) with steps of U=0.5 m, so that in total 200 steps steps are taken with a large number of iterations per step. Smaller values of step U do not yield significantly different results.

Table 2 gives the values of height F_S* and gradient G_S*/p, at the same selected values of distance X of Table 1 and 2, using Equations 2.7 and 2.8 (i.e. ignoring the energy balance). It is seen from Table 2.2 that the Fn* value (i.e the value of F* at X=N=100 m) equals 4.142 m. This is in agreement with the value Fn*=4.142 m that can be calculated directly from Equation 2.4.

Comparison of the tables learns that the Fn value (i.e. the value of F at X=N=100 m) of Table 1 (Fn=2.972) is considerably smaller than the Fn* value (4.142 m) of Table 2 (i.e. without energy balance). This is also shown in Figure 3.

REFERENCE

Wesseling, J. (1973). Subsurface flow into drains. In: Drainage Principles and Applications, Vol. II: Yheories of Filed Drainage and Watershed Runoff, p. 1-56. Publ 16, ILRI, Wageningen, The Netherlands.

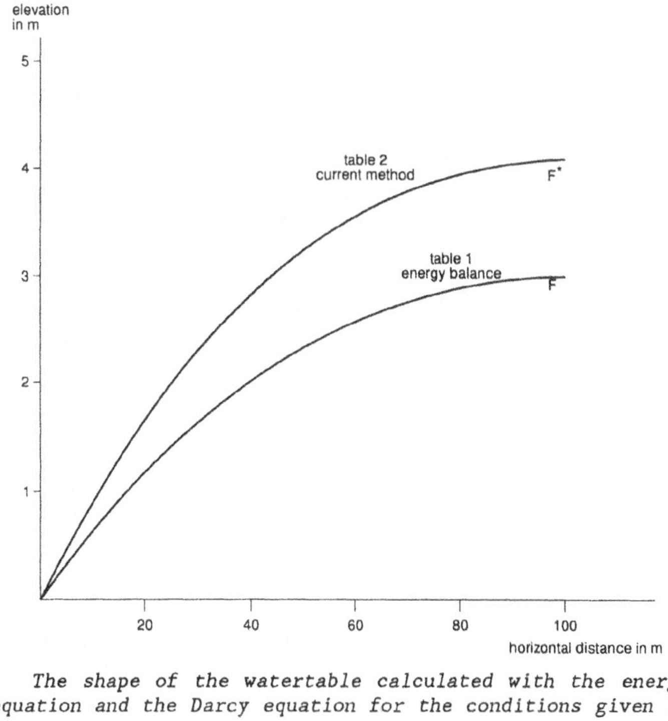

Figure 3. The shape of the watertable calculated with the energy balance equation and the Darcy equation for the conditions given in the example.

Table 1. Results of the calculations of the height of the watertable using a numerical and iterative solution of the hydraulic energy balance with steps U=0.5 m for the conditions described the example, using Equations 2.5 and 2.6.

Distance X (m)	Height of the water-table F (m)	Gradient of F G/U (m/m)	Gradient needed for the flow A (m/m)	Adjustment of A due to the energy of recharge B (m/m)
1	0.070	0.069	0.099	-0.029
2	0.138	0.068	0.097	-0.029
5	0.336	0.064	0.092	-0.028
10	0.643	0.059	0.085	-0.026
20	1.181	0.049	0.072	-0.022
30	1.630	0.041	0.060	-0.019
40	2.004	0.034	0.050	-0.016
50	2.310	0.027	0.041	-0.013
60	2.553	0.021	0.032	-0.011
70	2.739	0.016	0.024	-0.008
80	2.869	0.010	0.016	-0.005
90	2.947	0.005	0.008	-0.003
95	2.966	0.003	0.004	-0.001
98	2.971	0.001	0.002	-0.001
99	2.972	0.001	0.001	-0.000
100 (N)	2.972	0.000	0.000	-0.000

Table 2. Results of the calculations of the level of the watertable using a numerical solution of Equation 2.7 and 2.8 (i.e. without energy balance), with steps U=0.5 m, for the conditions described in the example.

Distance X (m)	Height of the watertable F* (m)	Gradient of F* G*/U (m/m)
1	0.099	0.099
2	0.196	0.097
5	0.476	0.092
10	0.909	0.083
20	1.662	0.069
30	2.288	0.058
40	2.806	0.047
50	3.229	0.038
60	3.564	0.030
70	3.820	0.022
80	4.000	0.015
90	4.107	0.007
95	4.133	0.004
98	4.141	0.002
99	4.142	0.001
100 (N)	4.142	0.000

ANALYSIS OF PUMP TEST DATA ON KASAI RIVER BED (WEST BENGAL) : A CASE STUDY

H. P. Patra, M. Bandopadhyay, G. Banerjee, S. N. Mukherjee
Indian Institute of Technology, Kharagpur, (W.B.)

and

M. Chowdhury

Public Health Engineering Directorate (W.B)

ABSTRACT : Pump tests were carried out on Kasai river bed in connection with augmentation of existing water supply of Kharagpur Municipality in the district of Midnapore, West Bengal to cater the need of a population of more than two lacs of the municipal township and for the growing demand of the industries. The draw down data collected during pump tests were analysed to determine the aquifer characteristics, namely, the storage co-efficient (S), transmissibility co-efficient (T). Methods adopted were Jacob's and modified Theis for un-confined condition and Hantush inflection point method for semi-confined condition. The values of S and T as obtained from the analysis of data are to be used subsequently for the design of radial collector wells in the area.

INTRODUCTION :

Kharagpur is one of the important railway towns of West Bengal situated at a distance of 115 Km away from Calcutta. The place is well connected by both rail and road with all parts of India. Some of the important manufacturing industries like Tata Bearings, Tata Davy, Scooter India Ltd. etc. are in operation in full swing within the proximity of the town. Besides the above, Institute like I.I.T., Kharagpur and Vidyasagar University are located close to the town. More new industries are expected to come for setting their units at Nimpura, the industrial zone earmarked by the Government of West Bengal in close vicinity of Railway station.

It has been projected that the future population of Kharagpur Municipality would be around 2.0 lacs for the year 2011. The total water demand for the year 2011 comes to 27 MLD considering per capita demand as 135 L/D.

In addition to above, industrial water demand, as per West Bengal Industrial infrastructural Development Corporation (WBIDC), for setting up an industrial growth centre at Nimpura is calculated as 5.5 MLD. Hence, total water demand comes to 32.5 MLD for the area.

V. P. Singh and B. Kumar (eds.), Subsurface-Water Hydrology, 161–172.

At present, only a part of the municipal area is provided with piped water supply through deep tube wells. Approximately, a total quantity of 5.0 MLD of water is available in existing municipal water supply system.

In view of necessary augmentation of the present municipality water supply a detailed investigation was carried out considering the subsurface water of the river Kasai as the prime source. It has been also proposed that the system of collecting the water would be through radial collector wells.

In order to study the aquifer properties of the proposed site, for designing of the radial collector wells and to asses the yield of well, several bore holes were sunk and pumping tests carried out during the lean season.

PUMPING TEST :

The location of boreholes for the pumping test on Kasai river bed for the present work is shown in Fig.1, where the positions of both pumping and observation wells are shown. The wells were pumped at a constant rate of 1135 lpm on different days for periods ranging up to 24 hours. Draw down and recovery data were collected in the observation wells. The draw down data collected were analysed to find out the aquifer parameters, e.g., storage coefficient (S) and transmissibility coefficient (T).

ANALYSIS OF PUMP TEST DATA :

The lithological logs of the boreholes in the study area shown in Fig.2 indicated the existence of clay lenses scattered in the subsurface with a variation in thickness. Since the lenticular clays are not found to be interconnected, the concept of unconfined condition was first taken into consideration and S and T values were computed using (i) Jacob's straight line method and (ii) modified Theis method which are discussed below.

(i) Jacob's straight line method :

It is known that Theis non-equilibrium formula for determining the hydraulic properties T and S of an aquifer can be given as (Raghunath, 1987) :

$$S = \frac{Q}{4\pi T} W(u) \qquad (1)$$

and

$$T = \frac{2.3Q}{4\pi \ \Delta s} \qquad (2)$$

with usual notations.

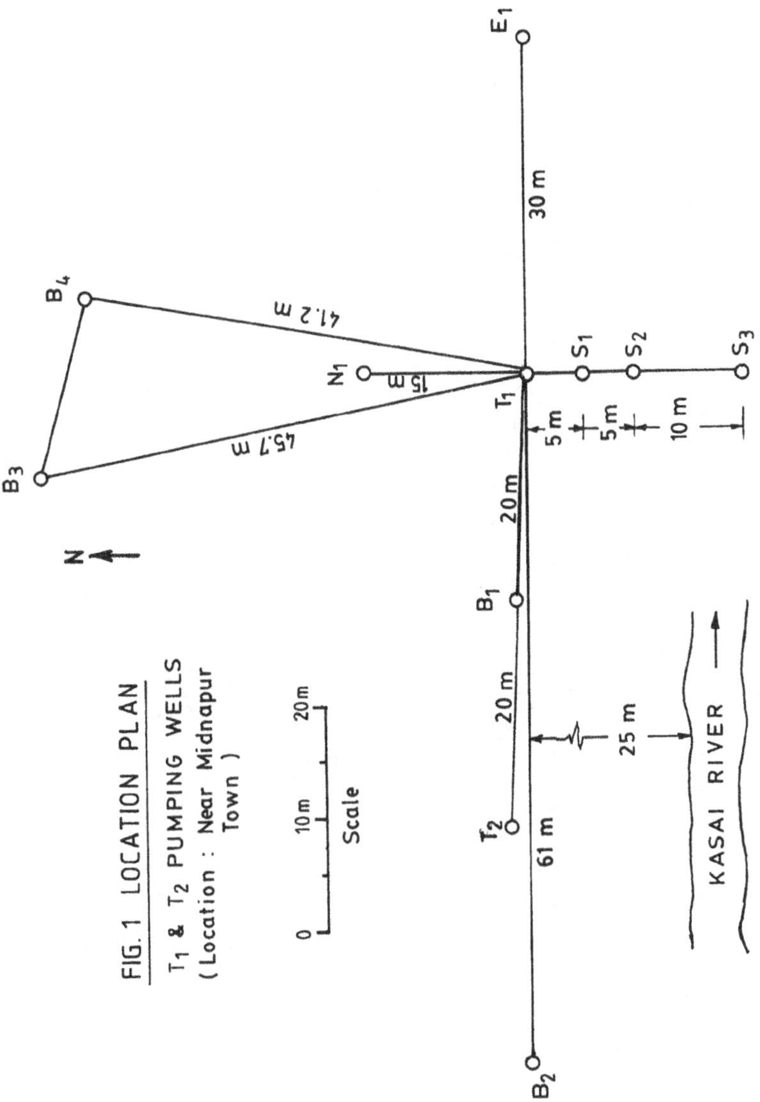

FIG. 1 LOCATION PLAN

T_1 & T_2 PUMPING WELLS
(Location : Near Midnapur Town)

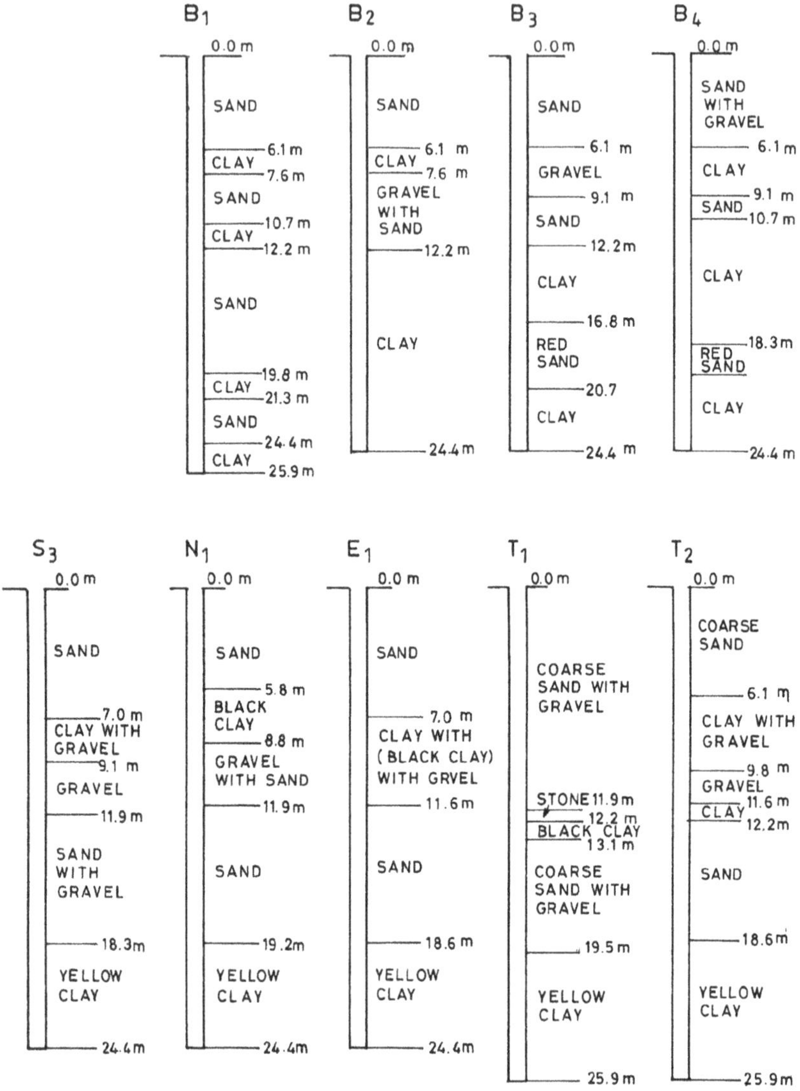

FIG. 2 LAYER CHARTS
(NOT TO SCALE)

By extrapolating the straight line of the semilog plot to intersect the zero-draw down axis at to, the time for s = 0, can be noted and S can be computed as

$$S = \frac{2.25\ T\ t_o}{r^2} \qquad (3)$$

The above equations are used to compute T and S (Table-I) with the sets of data from the area under study by Jacob's method for pumping wells T_1 and T_2 (Figs. 3 & 4).

(ii) Modified Theis method :

In this method a composite drawdown graph is drawn by plotting s vs. t/r^2 on a semi-log paper for all the observation wells (Figs. 5-6).

Δs per log cycle of t/r^2 is measured and aquifer constants T and S are determined from standard equations (Raghunath, 1987) :

$$T = \frac{2.3\ Q}{4\pi\ \Delta s} \quad \text{and} \quad S = 2.25T\ [t/r^2]_o \qquad (4)$$

The values obtained by (Figs. 5-6) assuming the aquifer as unconfined are presented in Table-I. The values of S presented in Table-I appeared rather low possibly due to scattered clay lenses. Under the circumstances, the aquifer is assumed as a semi-confined one and analysis of the data was made through Hantush inflection point method.

Hantus inflection point method :

Hantush and Jacob (1955) and Hantus (1956,1964) have given the following equation for steady - state draw down in a leaky artesian aquifer :

$$s = \frac{Q}{2\pi T} K_o\ (r/B) \qquad (5)$$

with usual notations.

From the above, Hantush developed a method for determining T, S from the time - draw down data by reading on the plot (s vs. log t) the values of s_i, t_i, Δs_i where i refers to inflection point, i.e., the point where the draw down (s_i) is one-half of the final or equilibrium draw down. In this method the draw down measured in an observation well during pumping is plotted against time on a semilog paper and the data is extrapolated until the maximum draw down s_m is reached. The inflection point s_i is

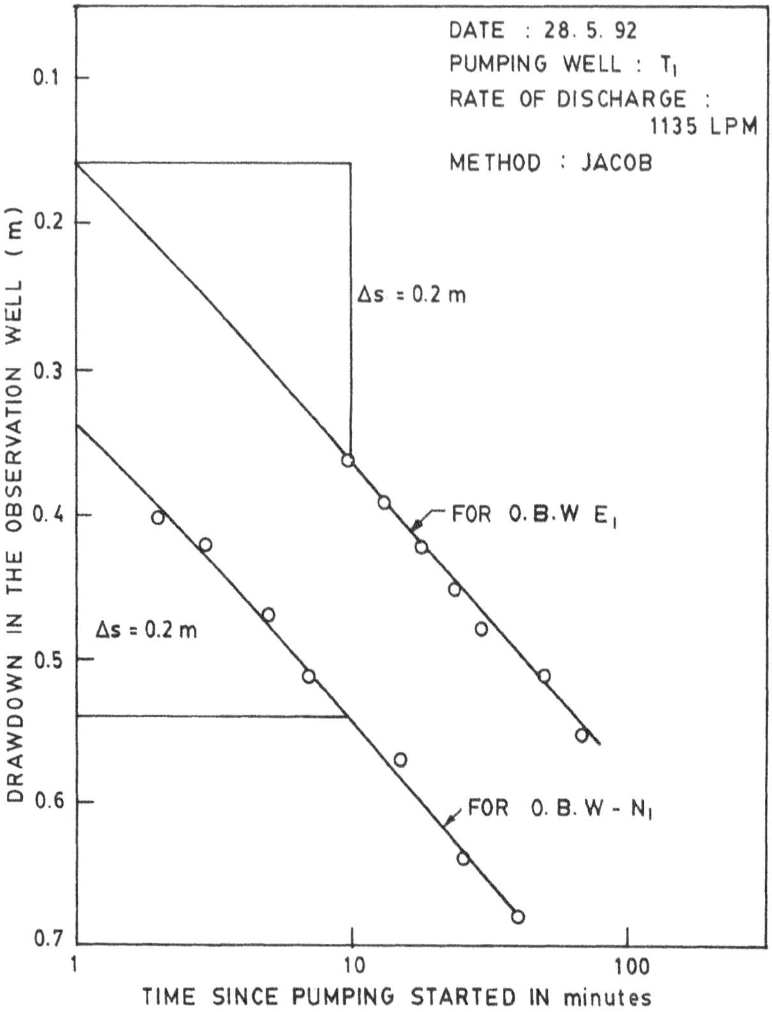
DATE : 28. 5. 92
PUMPING WELL : T1
RATE OF DISCHARGE : 1135 LPM
METHOD : JACOB
Δs = 0.2 m
FOR O.B.W E1
Δs = 0.2 m
FOR O. B. W - N1
DRAWDOWN IN THE OBSERVATION WELL (m)
0.1
0.2
0.3
0.4
0.5
0.6
0.7
1
10
100
TIME SINCE PUMPING STARTED IN minutes

FIG. 3

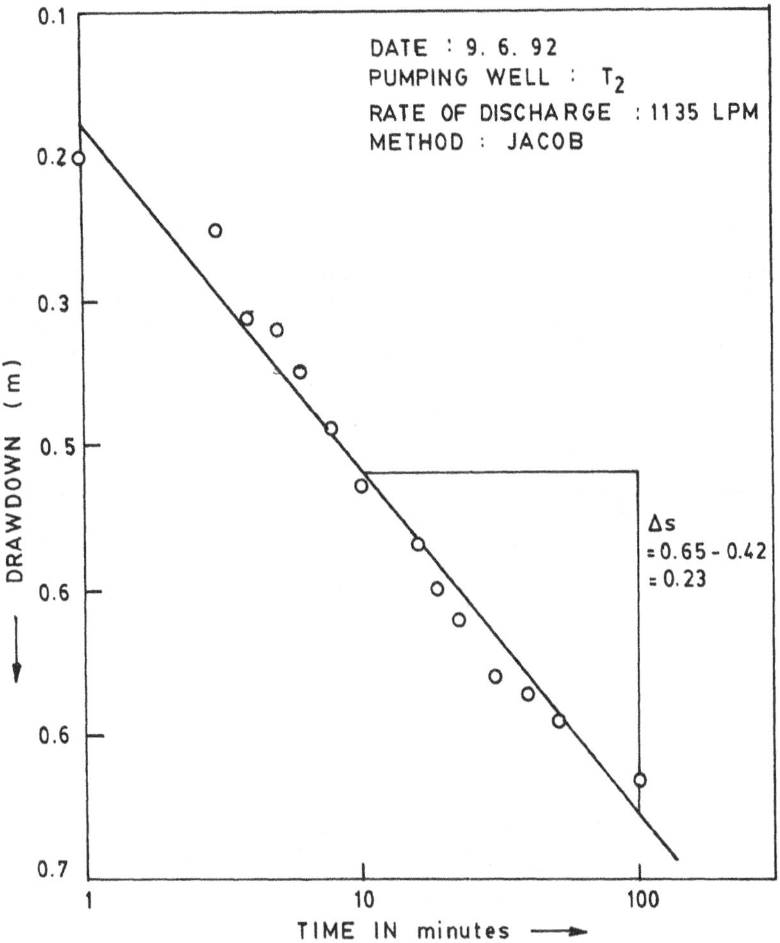
DATE : 9. 6. 92
PUMPING WELL : T2
RATE OF DISCHARGE : 1135 LPM
METHOD : JACOB
Δs
= 0.65 - 0.42
= 0.23
0.1
0.2
0.3
0.5
0.6
0.6
0.7
DRAWDOWN (m)
1
10
100
TIME IN minutes

FIG. 4

located on the draw down curve by taking $s_i = s_m/2$, where s_i = draw down at the inflection point. The slope of the draw down curve at the inflection point is determined graphically and t_i is read corresponding to the inflection points.

The values of T and S derived from these equations are given in Table - II and data plots shown on Fig. 7, for example.

RESULTS :

Pump test results are presented in Tables-I & II. For analysis of data, different graphs are plotted which are shown in Figures 3 to 7. Results of Jacob's and modified Theis methods are presented in Table - I while those of Hantush inflection point method are presented in Table II.

TABLE - I
(Values of S and T)

Date	Pumping well	Condition assumed	Methods adopted	S x (10^{-4})	T m^2/day
22.5.92	T_2	Unconfined Aquifer	Jacob's method	2.74	1424.00
			Modified Theis method	1.86	1196.57
28.5.92	T_1	Unconfined Aquifer	Jacob's method	3.05	1095.60
			Modified Theis method	2.81	1150.00
09.6.92	T_2	Unconfined Aquifer	Jacob's method	2.00	1300.00
			Modified Theis method	2.12	1359.27

TABLE - II
(Values of S and T)

Date	Pumping well	Condition assumed	Method adopted	S x (10^{-4})	T m^2/day
22.5.92	T_2	Semi-confined aquifer	Hantush method	1.395	1239.40
28.5.92	T_1	-do-	-do-	3.940	1328.00
09.6.92	T_2	-do-	-do-	3.260	1235.58

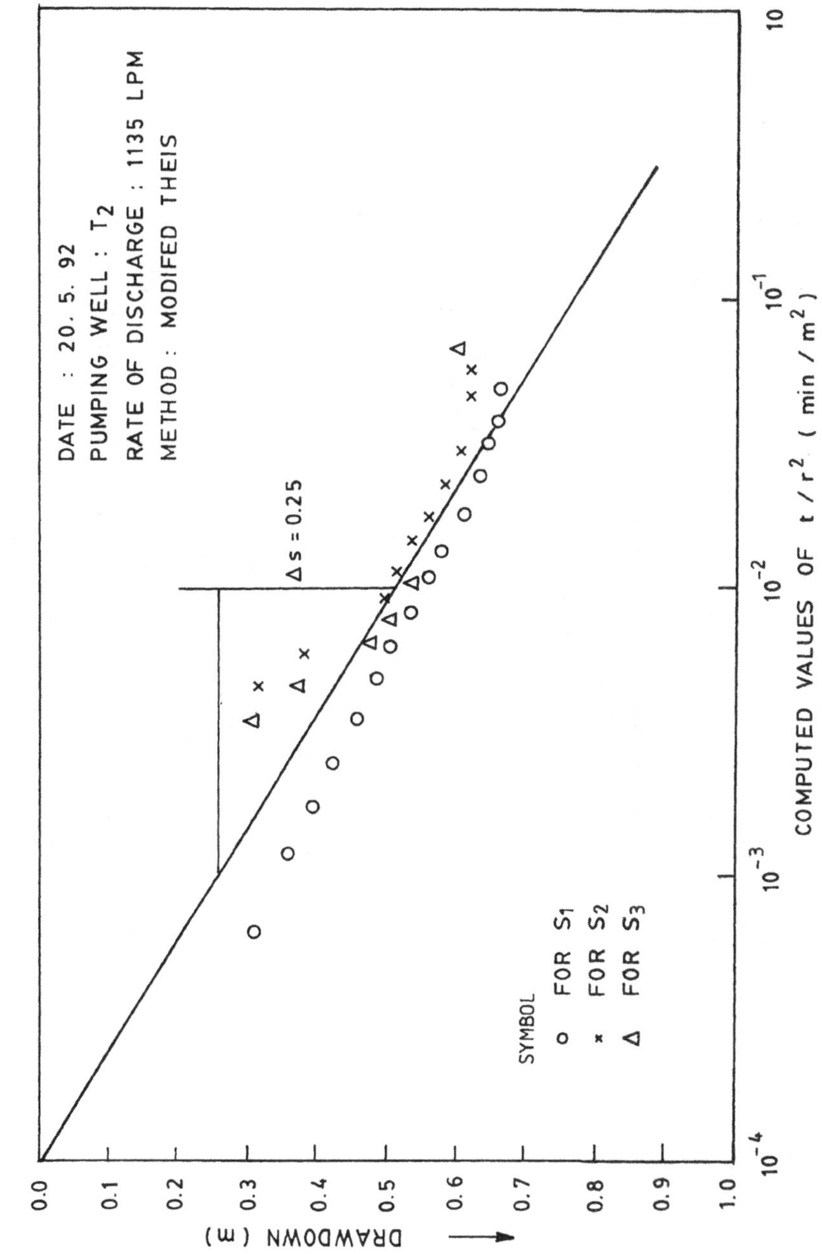
DATE : 20. 5. 92
PUMPING WELL : T_2
RATE OF DISCHARGE : 1135 LPM
METHOD : MODIFED THEIS
Δs = 0.25
SYMBOL
o FOR S_1
x FOR S_2
Δ FOR S_3
DRAWDOWN (m)
0.0
0.1
0.2
0.3
0.4
0.5
0.6
0.7
0.8
0.9
1.0
10^{-4}
10^{-3}
10^{-2}
10^{-1}
10
COMPUTED VALUES OF t / r^2 (min / m^2)

FIG. 5

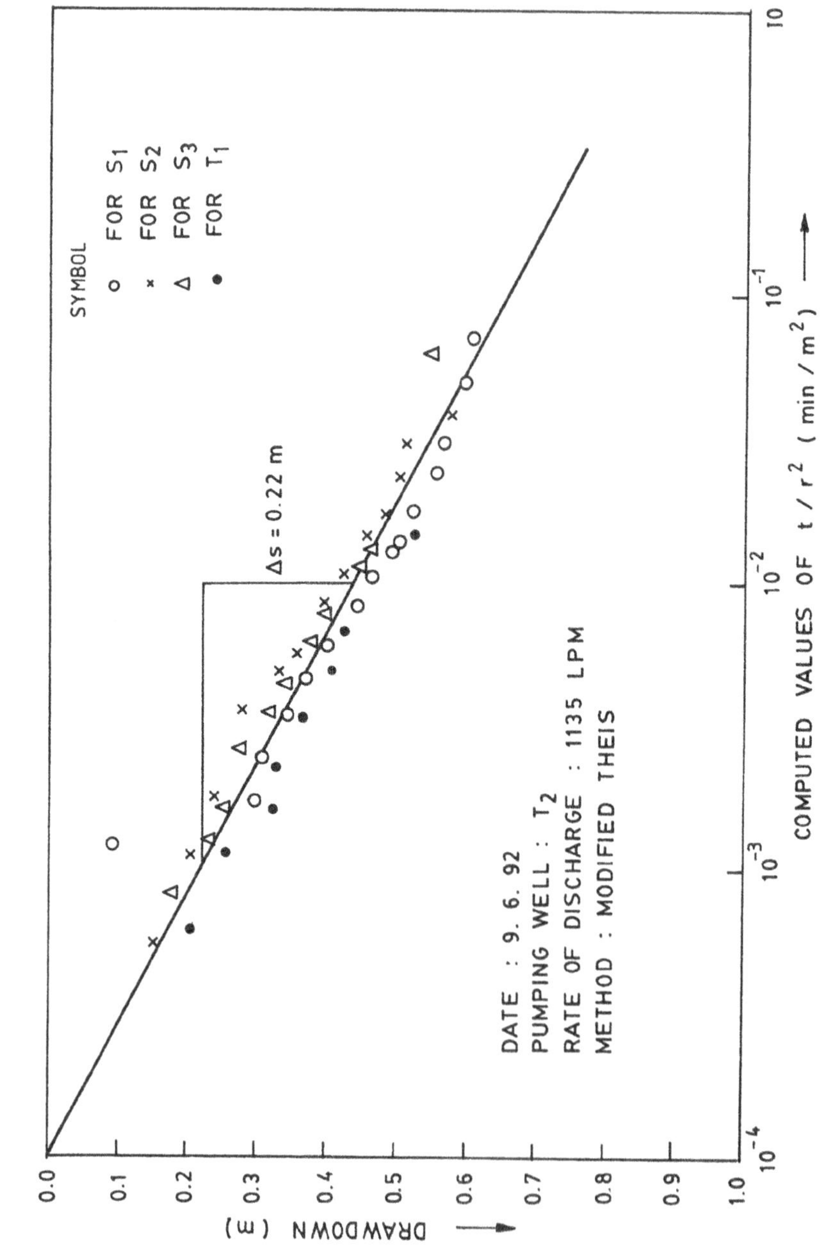
SYMBOL
o FOR S1
× FOR S2
Δ FOR S3
• FOR T1
Δs = 0.22 m
DATE : 9. 6. 92
PUMPING WELL : T2
RATE OF DISCHARGE : 1135 LPM
METHOD : MODIFIED THEIS
DRAWDOWN (m)
0.0
0.1
0.2
0.3
0.4
0.5
0.6
0.7
0.8
0.9
1.0
10^-4
10^-3
10^-2
10^-1
10
COMPUTED VALUES OF t / r^2 (min / m^2)

FIG. 6

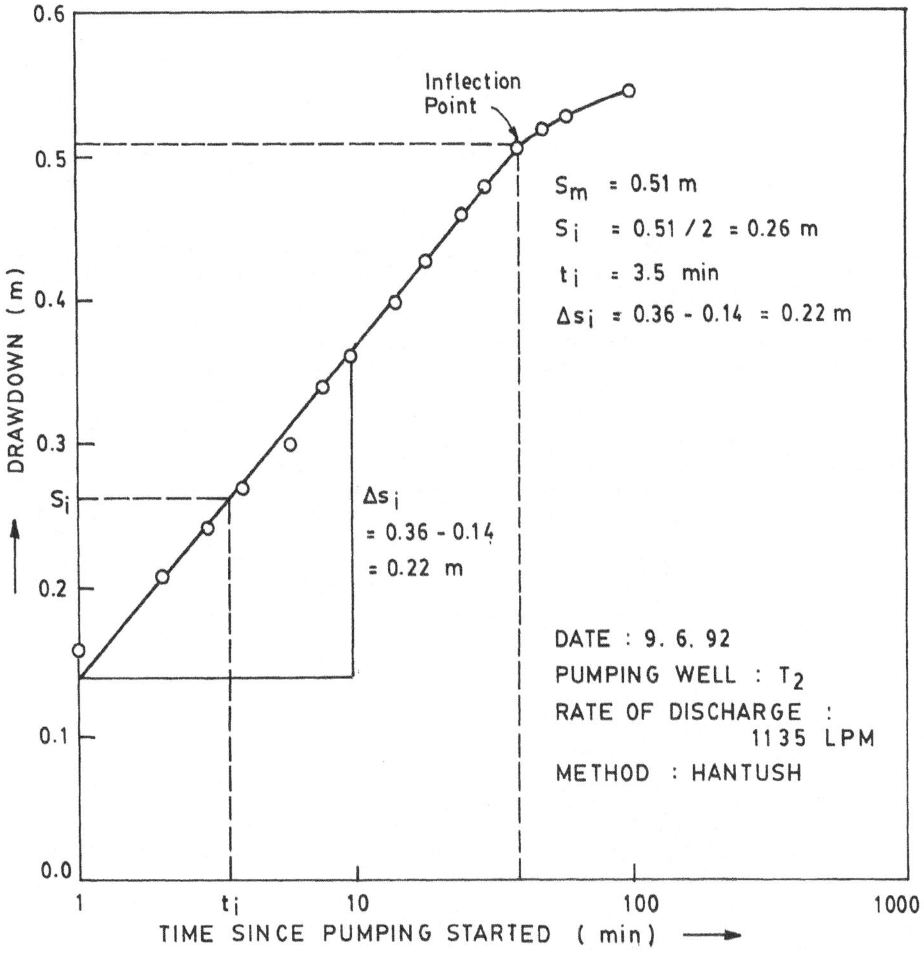
0.6
0.5
0.4
0.3
0.2
0.1
0.0
DRAWDOWN (m)
Inflection Point
Sm = 0.51 m
Si = 0.51 / 2 = 0.26 m
ti = 3.5 min
Δsi = 0.36 - 0.14 = 0.22 m
Si
Δsi
= 0.36 - 0.14
= 0.22 m
DATE : 9. 6. 92
PUMPING WELL : T2
RATE OF DISCHARGE :
1135 LPM
METHOD : HANTUSH
1
ti
10
100
1000
TIME SINCE PUMPING STARTED (min)

FIG. 7

DISCUSSION ON RESULTS :

The values of S and T obtained by Jacob's method and modified Theis method are presented in Table - I assuming the aquifer as unconfined. According to Raghunath (1987), the value of S for the unconfined aquifer lies normally in the range of 0.05 to 0.30 whereas for confined aquifer the S value range is 5×10^{-5} to 5×10^{-3}.

S values obtained, assuming the aquifer as unconfined, by Jacob's and Theis methods (Table - I) were found to be in the range of 1.9×10^{-4} to 3.05×10^{-4}, which does not indicate the aquifer to be unconfined one. Secondly, the values agreed well when the pump test data were analysed by Hantush inflection point method by assuming confined condition. The values of S obtained were in the range of 1.39×10^{-4} to 3.94×10^{-4} which are indicative of a semiconfined aquifer. From the lithological charts (Fig. 2) it was evident that clay lenses were present within the water bearing formations affecting S value. However, average transmissibility coefficient T estimated as 1400 m^2/day was found to be satisfactory. Further work on pump test on the Kasai river bank is in progress for the design of radial collector wells.

REFERENCES :

1. Raghunath, M.H., Ground Water, Wiley Eastern Ltd., New Delhi, 1987.

2. Hantush, M.S. and Jacob, C.E., Non steady radial flow in an infinite leaky aquifier, American Geophys. Un. Trans. 36, 95-100, 1955.

3. Hantush, M.S., Analysis of data from Tests in Leaky Aquifier Trans, Am. Geophys Un. Vol. 37, 1956.

4. Hantush, M.S., Advances in Hydroscience Vol.I, Academic Press, New York, 1964.

SECTION 5:

GROUNDWATER RESOURCE EVALUATION

GROUNDWATER RESOURCES EVALUATION

Ashim Das Gupta
Professor, Water Resources Engineering
Asian Institute of Technology
P.O. Box 2754, Bangkok 10501
Thailand

ABSTRACT: Evaluation of long-term natural recharge based on water balance method combined with the evaluation of dynamic response of aquifer system provide an appropriate tool for assessing the long-term sustainable yield of the groundwater system. This approach has been elaborated with reference to two case studies: one for the groundwater basin in the Kathmandu Valley of Nepal and the other for the coastal aquifer of Mannar Island in Sri Lanka.

1. Introduction

The development and management of groundwater requires an estimate of the sustained yield of the aquifer. Although it is usually limited by local aquifer transmissivity and storativity, the current rate of recharge can become a practical as well as theoretical limit to the sustained yield. Estimating the rate of aquifer replenishment is probably the most difficult of all measures in the evaluation of groundwater resources. Groundwater recharge is an intermittent and irregular process in the hydrological cycle. Direct assessment of the component of recharge from rainfall on a regional scale is difficult. Several indirect approaches, however, can be used to estimate the groundwater recharge. These include the hydrologic budget method, conventional water balance method, hydrograph separation method, tracer technique, and conventional modeling of the groundwater flow system. The conventional water balance method on long-term basis is the most widely used approach for estimating groundwater recharge. The hydrologic budget method has the advantage of relating groundwater recharge to other components of the hydrologic cycle and thus providing a further understanding of the hydrology in the broader context.

Most of these modeling approaches need data that are not readily available. However, for planning purpose, a preliminary estimate is needed. Depending on the physical system, a particular method or a combination of approaches is to be adopted seeking an appropriate balance between complexity and accuracy considering the quality

V. P. Singh and B. Kumar (eds.), Subsurface-Water Hydrology, 175–191.

of available data and the needs for practical design and analysis. This aspect has been addressed to in this paper with reference to two case studies: one for the groundwater basin in the Kathmandu Valley of Nepal and the other for the coastal aquifer of Mannar Island in Sri Lanka. The recharge to the groundwater basin in Kathmandu Valley is estimated by a water balance model and compared with estimates by other indirect methods. A simulation model is then used to assess the long-term sustainable withdrawal restricting the extent of drawdown. The water balance estimates in combination with the simulation of movement of fresh-/salt-water interface have been used to assess the natural recharge to aquifer underlying Mannar Island of Sri Lanka. The long-term level of safe groundwater extraction is then assessed by limiting the upconing effect.

2. Assessment for Kathmandu Valley, Nepal

2.1 STUDY AREA

The Kathmandu valley is more or less circular in configuration and has a total area of 899 km^2. The topographic features comprise of gently sloping valley floor, valley plain terrace with scarp faces, screes and talus cone deposition, together with the flood plain. The valley is filled with quaternary lacustrine and fluvial sediments comprising of gravel, sand, silt and clay with lignite seams. These sediments lie unconformably on the paleozoic and pre-cambrian rock. Based on physical and geological structures the groundwater basin area of 327 km^2 may be divided into three groundwater districts: northern district with area 157 km^2, central district with area 114.5 km^2 and southern district with area 55.5 km^2 (Figure 1).

The climate of the area is warm temperate semitropical and is largely affected by the prevailing monsoon in the region. The mean relative humidity is 75% and wind velocity rises (till May) upto an average of 1.8 m/s and decreases after monsoon season until December to an average of 0.5 m/s. The valley is situated along the upstream reaches of the Bagmati river. The watershed of the study area has an approximately circular shape with area of 585 km^2 at outlet of the basin at Chovar. The Bagmati river is the only river which drains water collected in the valley to the south and dissects the mountain of Mahabharat range to the southwest of the valley. There are nine major tributaries joining the main stream in the valley. About 50% of the total water demand in the valley is fulfilled from groundwater. There are seven major springs at the base of the mountains gushing from the faulted limestone layers. There are 31 spouts with discharge ranging from 3 to 5 ℓ/s in the dry season and 5 to 15 ℓ/s in the rainy season. Groundwater extraction is more during winter season due to less discharge available in river compared to summer season. The total rechargeable area in the basin is 86 km^2, about 69% of this lying in the northern district, 7% on the central and 24% on the southern district. The hydraulic conductivity of the unconsolidated formation ranges from 10^{-4} cm/s to 10^{-2} cm/s. The total groundwater abstraction estimated in 1989 was 45,500 m^3/d out of which 6,330 m^3/d was from private wells.

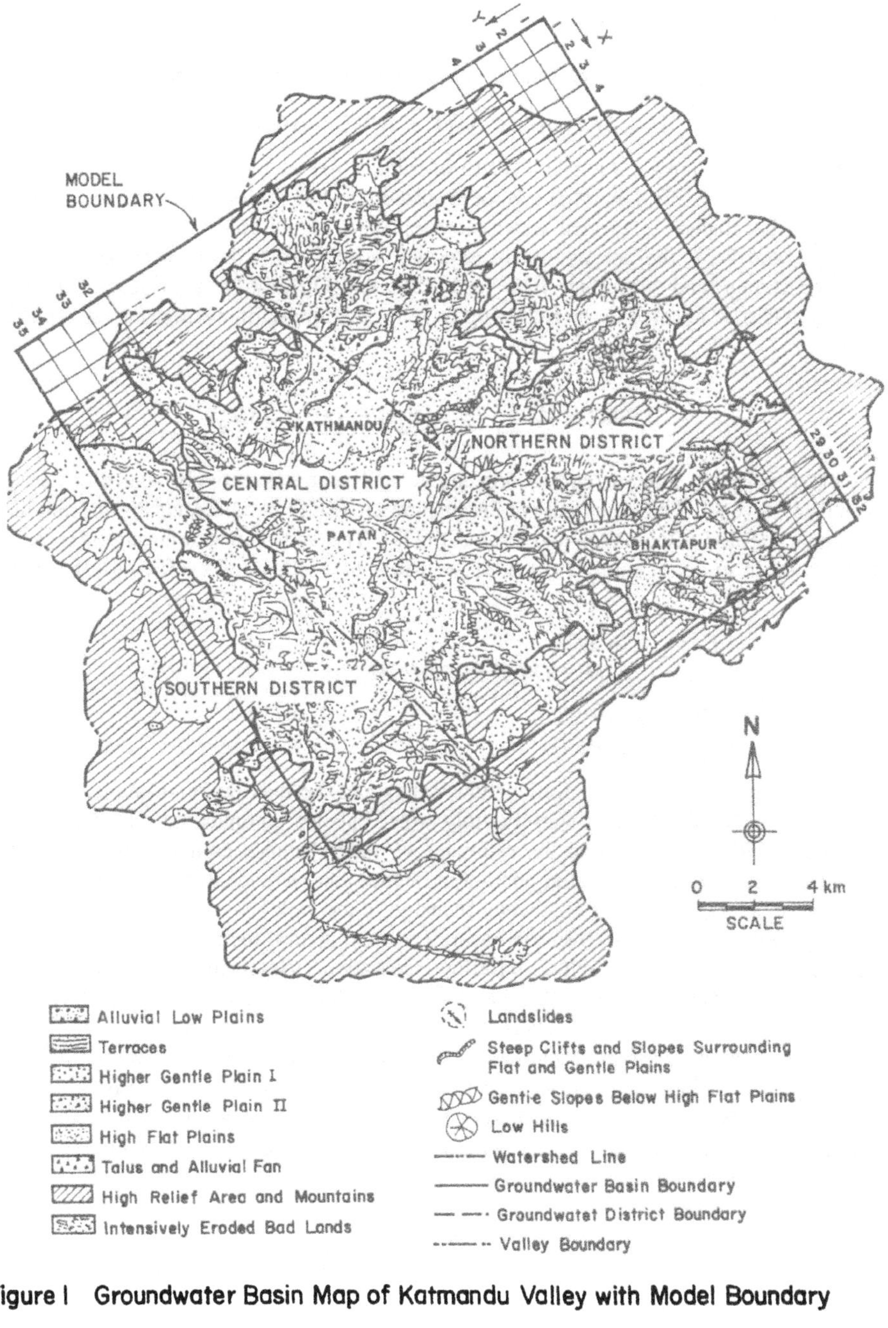

Figure I Groundwater Basin Map of Katmandu Valley with Model Boundary

2.2 RECHARGE

The water balance model representing mass conservation can be expressed as

$$P = SF + AE + \Delta SM + \Delta SW + \Delta GW + Q_o \tag{1}$$

where P is the equivalent uniform depth of rainfall over a basin; SF is the stream flow from the basin; AE is the actual evapotranspiration; ΔSM is the change in soil moisture; ΔSW is the change in surface water storage; ΔGW is the change in the groundwater storage; and Q_o is the net amount of exported water. For a basin with no significant surface storage and subsurface flow loss, ΔSW and Q_o may be neglected. The above equation is then reduced to the following form for the long-term average annual recharge, GRC

$$GRC = \frac{1}{n} \sum_{j=1}^{n} \sum_{i=1}^{12} (P_{ij} - SF_{ij} - AE_{ij} - \Delta SM_{ij}) \tag{2}$$

where i and j are indices representing month and year, respectively; and n is the total number of year. The potential evapotranspiration (PET) is calculated by the modified Penman method (Doorenbos and Pruitt, 1977). The actual evapotranspiration (AE) is expressed in terms of rainfall (P), PET, soil moisture (SM) and coefficient of land use (C_{land}).

$$\begin{aligned} &For \quad P < PET \\ &AE = C_{land}\,[P + (PET\text{-}P)\,(SM/SM_{max})] \end{aligned} \tag{3}$$

$$\begin{aligned} &For \quad P > PET \\ &AE = C_{land}\,PET \end{aligned} \tag{4}$$

C_{land} is estimated as 0.91 for Kathmandu Valley and the maximum soil moisture, SM_{max} is taken as 225 mm. The mean monthly water balance computation was done for the period 1976 to 1988 and the estimates of water balance components are given in Figure 2. For details of the computation, reference can be made to Shrestha (1990). The long-term average annual groundwater recharge to the valley was estimated to be 51 mm or 3 percent of the rainfall over the basin. Potential natural recharge was also computed based on other methods such as baseflow separation (Viessman et al., 1977), specific yield method (Chow, 1964), chloride balance method (Binnie & Partners, 1988) and groundwater flow method (McWhorter and Sunada, 1985). Comparison of recharge values estimated by different approaches is provided in Table 1. The average of annual recharge by various methods is found to be close to that calculated based on water balance computation. The estimated average annual recharge to the Kathmandu Valley basin is therefore taken as 51 mm per year which is equivalent to 45,690 m^3/d.

2.3 SUSTAINABLE YIELD

The long-term sustainable yield of the basin was evaluated through flow simulation. The

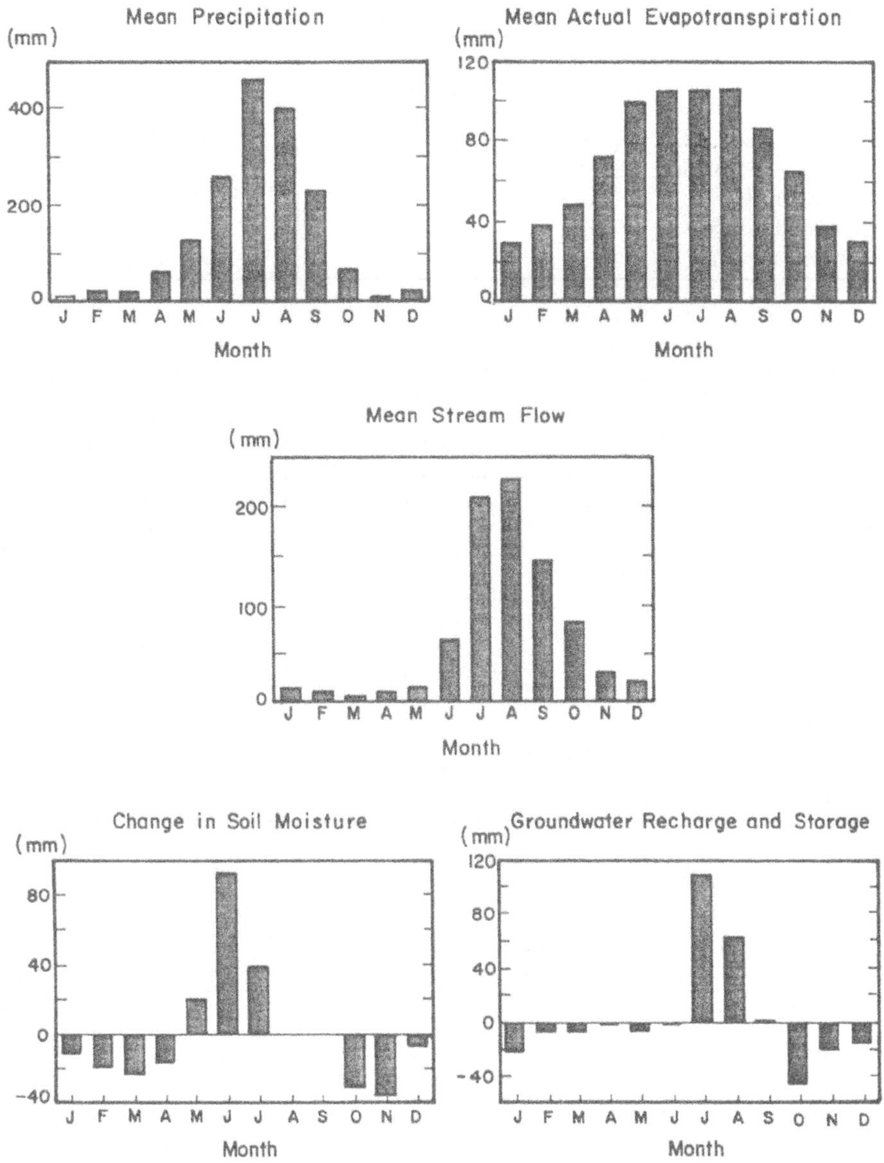

Figure 2 Mean Monthly Water Balance for Kathmandu Valley

Table 1 Comparison of recharge estimation for Kathmandu Valley by various methods

Methods	Recharge (mm per year)	Recharge (m^3/d)
Water Balance	51	45690
Base Flow Separation	55	49140
Specific Yield	38	33542
Chloride Balance	59	53606
Groundwater Flow	41	36394

Table 2 Estimation of sustainable yield for Kathmandu Valley

Year	Total Water Demand (x 1000) m^3/d	Groundwater Pumping	
		Sustainable Yield (m^3/d)	Percent of Total Demand
1990	100	47520	48
1991	109	47520	44
1992	119	47520	40
1993	127	47520	37
1994	136	47520	35
1995	145	42336	29
1996	153	42336	28
1997	161	42336	26
1998	168	42336	25
1999	176	46656	27
2000	184	46656	25

Table 3 Selected trials for the analysis of interface profile in Mannar Island

Trial	FRO	F	C,mm	Recharge, mm									
				1972	1973	1974	1975	1976	1977	1978	1979	1980	1981
1	0.00	0.20	50	0	205	0	0	153	127	212	270	0	15
2	0.05	0.05	50	0	186	0	0	139	119	274	247	0	31
3	0.05	0.10	100	0	183	0	0	121	105	180	248	0	9
4	0.05	0.05	150	0	108	0	0	123	119	180	247	0	0
5	0.10	0.20	50	0	124	0	0	97	48	148	166	0	0
6	0.10	0.05	150	0	149	0	0	65	48	134	166	0	0
7	0.15	0.20	100	0	88	0	0	55	43	88	121	0	0

FRO: Surface runoff factor; F: Dry factor; C: Root constant

groundwater basin was considered as a confined aquifer system. The two-dimensional groundwater flow equation is expressed as

$$\frac{\partial}{\partial x}\left(T_x \frac{\partial h}{\partial x}\right) + \frac{\partial}{\partial y}\left(T_y \frac{\partial h}{\partial y}\right) = S\frac{\partial h}{\partial t} - N + Q \tag{5}$$

where h is the piezometric head in the aquifer; T_x and T_y are transmissivities in x and y directions respectively; S is the storage coefficient; N is the natural and/or artificial replenishment rate; Q is the extraction rate; and t is the time. Equation (5) was solved numerically using an iterative alternating direction implicit finite difference scheme. The flow domain was discretized into 1120 rectangular grids with variable nodal spacing and the modelled area is shown in Figure 1. Model grids covering the groundwater basin area and model pumping locations are shown in Figure 3. Basin boundaries at the foothills were assumed to be impermeable. The transmissivity values used were in the range of 32 to 1146 m^2/d and a storage coefficient of 0.01 was assumed. The natural recharge was considered at a constant rate of 194 mm per year on specified recharge area as equivalent to recharge estimate corresponding to the basin area.

The model was calibrated during 1975 to 1989. The initial condition was assumed to be corresponding to the observed piezometric head distribution in 1970 as there was no significant pumpage during 1970 to 1975. Initially a steady state simulation was done through adjustment of transmissivity values to get the model results agreeing closely with the observation in 1970 as shown in Figure 4(a). With historical pumpage during 1975 to 1989, spatial distribution of pumpage at existing well locations were adjusted till the calculated piezometric heads in selected years agreed with the observed values. A comparison of calculated piezometric head distribution with observations in 1989 is shown in Figure 4(b). Following calibration the model was used to estimate the sustainable yield based on two criteria: economic pumping and allowable drawdown. It was assumed that the water charge and the energy cost will be increased proportionately so that benefit from and cost for operation remain constant. The maximum allowable pumping lift and allowable drawdown were calculated as 37 m and 20 m, respectively. Starting with groundwater extraction from 1990 equal to 50% of the total water demand, the predicted drawdown was found to be increasing beyond the maximum allowable drawdown by the year 1995. Predictions were then made for different pumping scenarios and the pumping pattern which met the requirement of allowable drawdown was considered as giving the long-term sustainable yield from the basin. The yearly pumping rates of this scheme compared to total demand during 1990 to 2000 are given in Table 2. The long-term sustainable yield was found to be 45478 m^3/d compared well with the estimates based on water balance study.

3. Assessment for Mannar Island, Sri Lanka

3.1 STUDY AREA

Mannar Island is situated along the northwest coast of Sri Lanka, and the township is

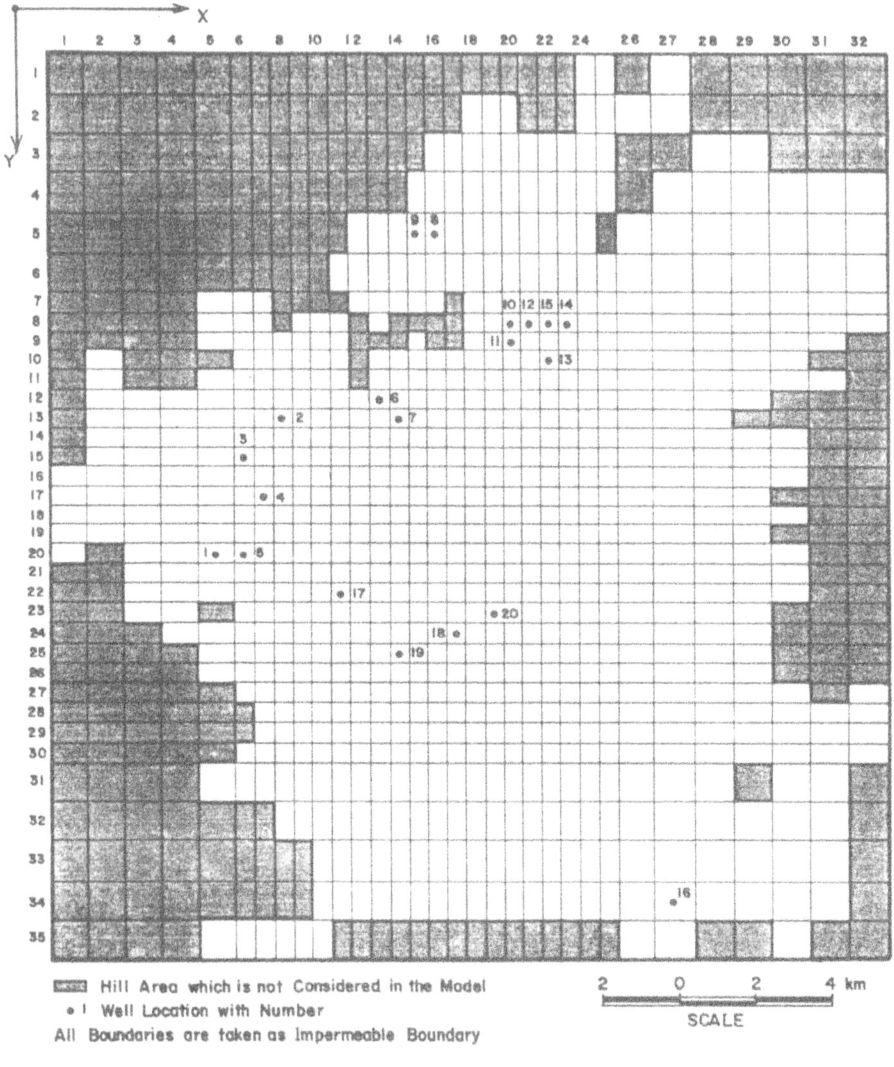

Figure 3 Model Grid with Well Location and Boundary Conditions

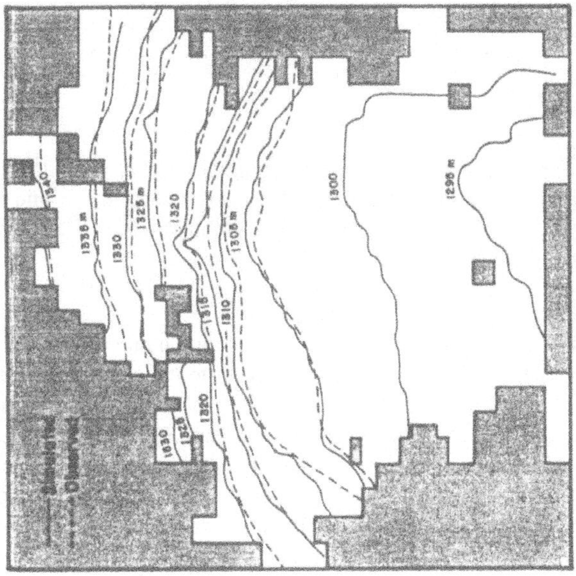

Figure 4 (a) Initial Piezometric Head Distribution in 1970 (Head Values above MSL)

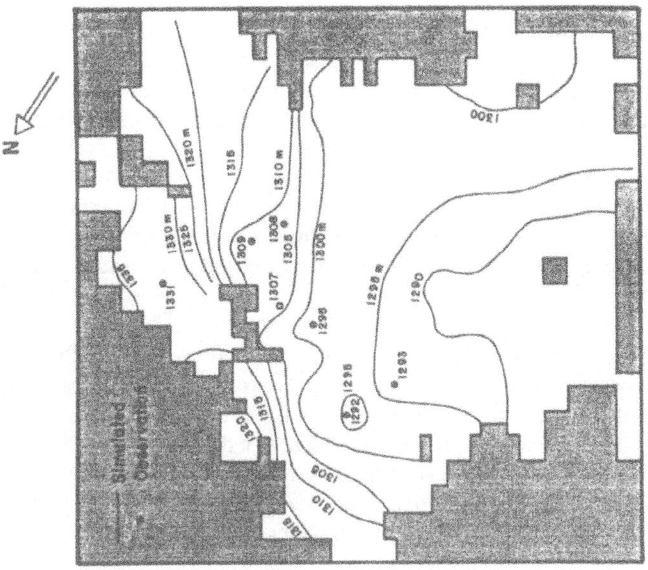

Figure 4 (b) Simulated Piezometric Head Distribution Pattern with Oserved Heads for 1989 (Head Values above MSL)

about 260 km north of Colombo. The extent of the island is 128 sq km and the study area is a narrow coastal belt which covers the villages of Thoddakadu, Sunny Village, and Thoddavelli (Figure 5). The land is generally flat except for the sand dune ridges rising along the axis of the elongated island. Lagoonal and windblown sand deposits are observed as two contrasting formations. This island has a semiarid climatic environment with three well-defined seasons: the wet season from October to December, an intermediate season from January to May, and the dry season from June to September. The average annual rainfall for 1972-1988 was 948 mm with the maximum monthly average being 235 mm for November. There is no surface-water storage in the island except for a brackish lagoon. The highly permeable nature of the land surface results in only minimal surface runoff.

The island is mainly covered by recent sediments of dune sand, alluvial, deltaic, and lagoonal deposits of quaternary age. The tertiary formations lie directly on the crystalline basement rocks. Lithological information (Rao et al., 1984) reveals the presence of sand and silt beds down to a depth of around 7 m, followed by calcareous sandstones, clays and siliceous limestone formations down to a depth of around 29 m, and crystalline rocks to a depth of 250 m. Most of the available well information was up to a depth of 30 m. Limited data on hydrogeologic characteristics indicate that the hydraulic conductivity of the shallow unconsolidated formation varies from 0.8×10^{-3} to 1.4×10^{-3} m/s and the specific yield value ranges from 0.07 to 0.20. The groundwater table in the study area is 0.25 to 0.75 m above mean sea level. The water supply in the area is generally derived from the available groundwater resource of a fresh-water lens floating over saline water. Groundwater is pumped from seven well fields (Figure 5), each with a cluster of three wells.

3.2 WATER BALANCE

The general form of the water balance equation, Equation (1) was used to estimate the groundwater recharge with proper specification of hydrologic components and parameters according to the physical system under consideration. Mathematically, the soil moisture balance is represented by the basic equation

$$P = AE + SF + \Delta SM \qquad (6)$$

The degree to which the potential and actual evapotranspiration rates diverge is a function of soil and vegetation characteristics and is explained in terms of a root constant which is a measure of the water readily available within the root zone (Rushton and Ward, 1979). The soil loses moisture at a rate, equal to the potential evapotranspiration of that time from the point of field capacity, up to the point at which it reaches the root constant beyond which the soil evaporates at a fraction of potential evapotranspiration, and this fraction is called the drying factor. The change in soil moisture storage is then expressed as a known fraction of the potential change in storage as

$$\Delta SM = F\ (P - SF - PE) \qquad (7)$$

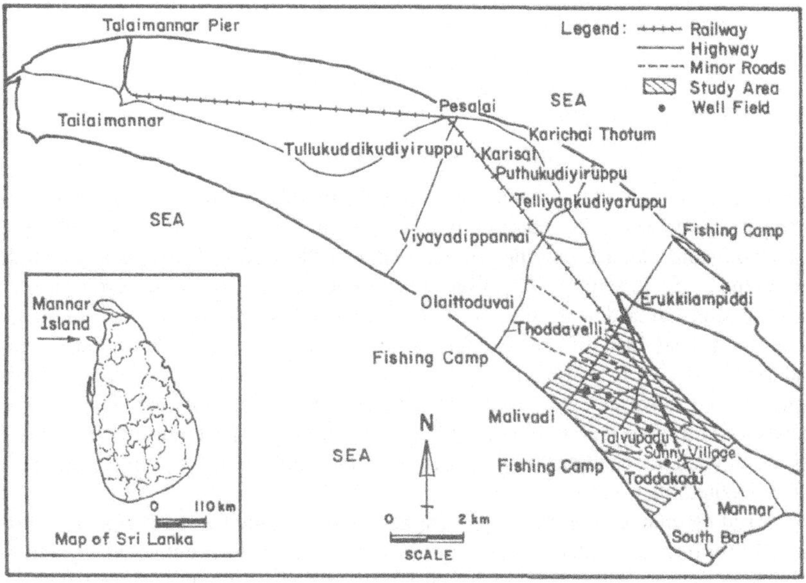

Figure 5 Location Map of the Study Area in Mannar Island, Sri Lanka

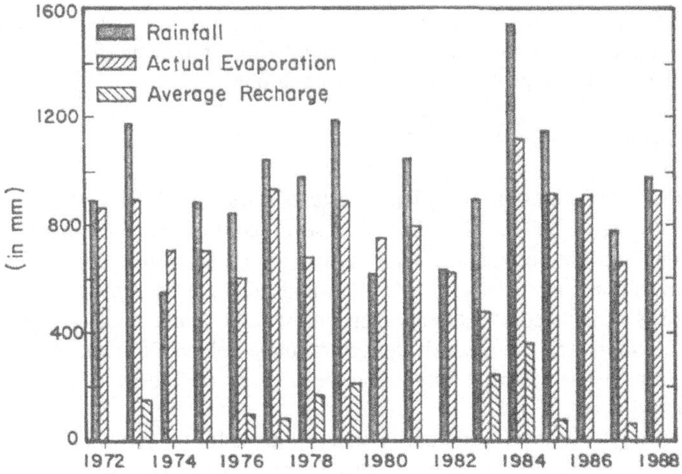

Figure 6 Annual Rainfall, Actual Evaporation and Average Annual Recharge Considering 36 Trial Estimates

where PE is the potential evapotranspiration; and F is the drying factors. Also, in terms of actual evapotranspiration, from Equation (6)

$$\Delta SM = P - SF - AE \tag{8}$$

From Equations (7) and (8)

$$AE = (1-F)\ (P-SF) + F.PE \tag{9}$$

Of the necessary input data for the estimation of groundwater recharge in the study area, only precipitation was measured directly while potential evapotranspiration, surface runoff and soil moisture balance parameters were either estimated by other means or assumed based on the physical setting. The monthly precipitation data for a meteorological station within the study area during the period 1972 to 1988 were used for the analysis. The surface runoff was insignificant except for some inland flow into the marshy areas. In the absence of any measurement and soil properties, the surface-water outflow was assumed to be a certain percentage of precipitation. Climatological data on monthly intervals for temperature, humidity, cloud amount and wind speed during the period 1972-1988 were used to calculate the reference evapotranspiration based on the modified Penman's equation.

Trial estimates of recharge (36 estimates) were made for the period 1972 to 1988 on a monthly basis considering the variations of runoff factor, root constant, and drying factor. The average annual recharge considering all trial estimates, annual rainfall, and annual actual evaporation are graphically shown in Figure 6. Recharge was insignificant during the years 1974, 1975, 1980, 1981, 1982 and 1986 (since annual rainfall during these periods was low), while high recharge estimates were obtained during the years 1979, 1983 and 1984. A set of seven trial estimates (Table 3) covering the range of variation of the parameters is selected as input for the analysis of the interface movement under recharge and pumping conditions. The probable set of parameters representing the field conditions is then found through the simulation of the interface profile and by comparing the calculated profile with the observed one.

3.3 FRESH-/SALT-WATER INTERFACE

Pumping from well fields in the Mannar Island started in 1981 and, from the field measurement taken at the end of 1981, three sectional profiles of the fresh-/ salt-water interface along sections running east-west across the island were prepared for the study area. The transition zone between fresh water and salt water is approximated as a sharp interface and is considered at a depth where the specific conductivity is between 2000-2500 micromhos/cm and the chloride concentration is below 250 ppm. The fresh-water lens extended to the sea front and the interface was assumed to have formed in a semi-infinite extent aquifer.

Hantush (1968) developed approximate solutions for the growth of the fresh-water lens under recharge conditions in an infinite aquifer. Selected solutions with appropriate modification were used to calculate the depth to the interface. As the depth of the fresh-

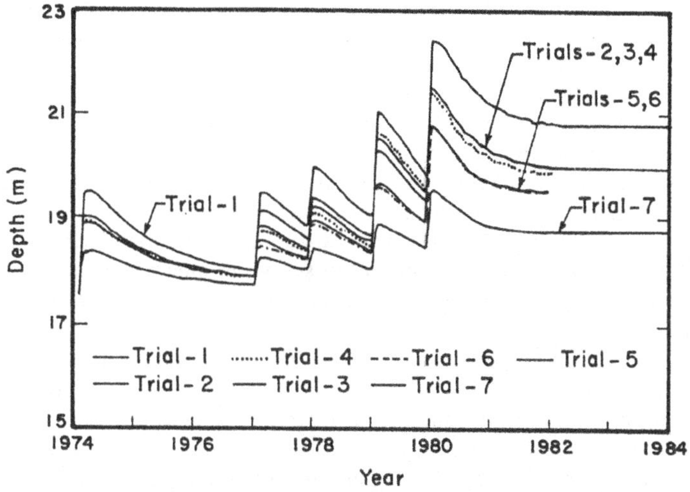

Figure 7 (a) Variation of Depth to Interface below MSL at the Center of the Section Due to Recharge for Selected Trials (Refer to Table 3)

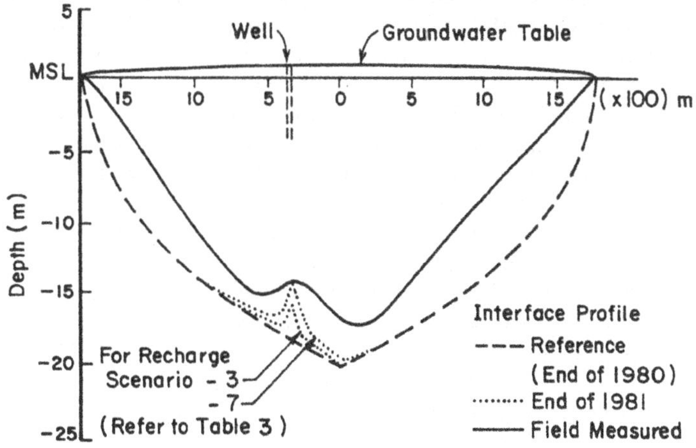

Figure 7 (b) Simulated Interface Profile at the End of 1981 for Different Recharge Scenarios Compared with Field Observed Profile

water flow at the sea front at any instant is small compared to the maximum thickness of the lens, the method of images was incorporated into the solutions for the infinite extent aquifer. Reference is made to the original paper for the detailed derivation of the solutions. The aquifer received recharge from rainfall and this resulted in the growth of the fresh-water lens. The decay of the interface was due to pumping and the resulting effect was superimposed on the growth of the interface profile to get the resultant fresh-/ salt-water interface profile under pumpage and recharge conditions. Also, when recharge ceases at any instant, the decay of the fresh-water lens was obtained by superimposing hypothetically on the flow system at that instant a rate of uniform discharge equal to that of the uniform recharge. The resultant fresh-/ salt-water interface is the superposition of the initial condition, the effects of vertical recharge, and the constant rate of pumping from wells.

The movement of the interface under pumpage and recharge conditions was simulated during the period 1972 to 1988. The observed interface profile at the end of 1981 was taken for calibrating the model response whereby the appropriate recharge sequence was ascertained. The pumping from Sunny Village well fields started in 1981 and the extraction rate was 102.3 m^3/d. The simulation of the fresh-/ salt-water interface during the period 1972-1988 was done in two steps. The flow condition prior to 1972 was assumed to be in a steady equilibrium condition and the initial interface profile was obtained based on solution given by Das Gupta and Gaikwad (1987).

The recharge calculated for the selected trials (Table 3) from the water balance study was used as the vertical percolation to the fresh-water lens. For the period 1972-1980, the aquifer was not stressed by pumping and the growth of the fresh-water lens occurred during those months when recharge occurred, while during other months the fresh-water lens decayed. The interface profiles were estimated during 1972-1980 by considering the natural recharge events of the selected trials given in Table 3. The corresponding variations of depth to the interface below MSL at the center of the section are shown in Figure 7(a). After 1980, due to the insignificant recharge of the lens, the profiles approached a steady state. Starting with the profiles at the end of 1980, the profile of the interface at the end of 1981 was simulated. The computed profiles for two recharge scenarios compared with the field observed one are shown in Figure 7(b). The estimated recharge sequence of trial 7 was found to be the best representation of the actual condition and this recharge sequence is shown diagrammatically in Figure 8. Details of analysis related to recharge estimations and simulation of interface movement are given in Amaraweera (1990) and Das Gupta and Amaraweera (1993). Following the recharge sequence of trial 7, the interface movement is then further simulated upto 1988 under pumpage and recharge condition. A continual upconing of interface was observed.

3.4 LONG-TERM SAFE WITHDRAWAL

The long-term safe withdrawal was assessed approximately through the predictive simulation of the interface upconing by limiting its extent at the well section to a safe level. Based on the experimental observations by Dagan and Bear (1968), the critical

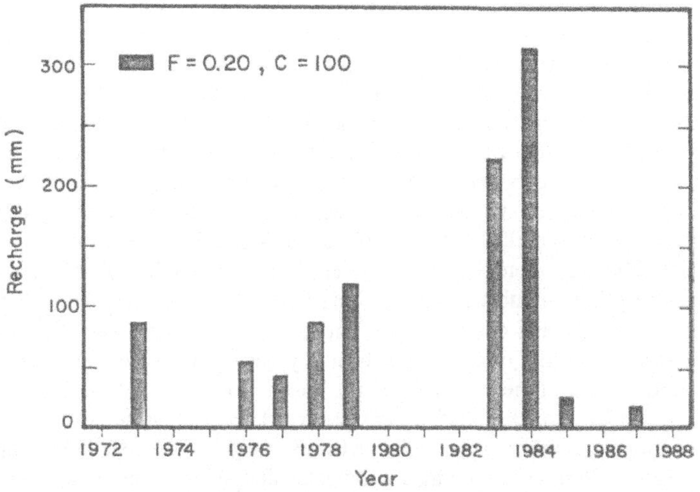

Figure 8 Representative Recharge Sequence Corresponding to the Agreement of Simulated Interface Profile with Field Observed One at the End of 1981

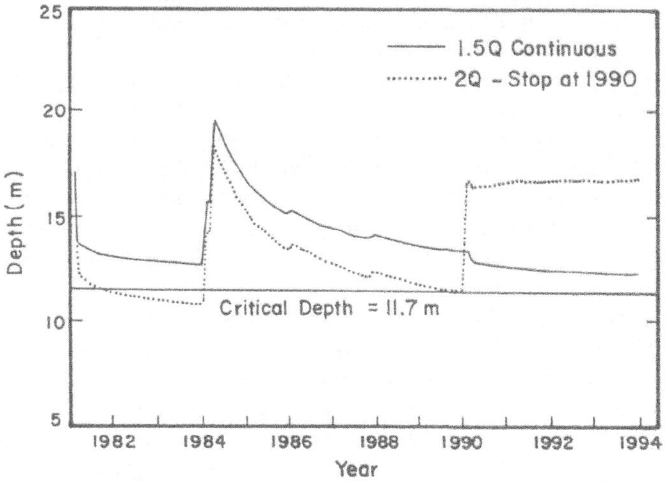

Figure 9 Predicted Depth to the Interface below MSL at the Well Section Due to Different Pumping Pattern During 1981 - 1993

upconing height is considered as one-third of the distance between the natural interface location and the bottom of the well. For the specific section, under consideration, by considering the depth of the well below MSL as 4 m, the interface location at the well section under natural flow condition be 15.53 m below MSL, and the critical upconing height, the depth to the interface at the well section below MSL in upconed state, should not be less than 11.7 m (defined as critical depth). The water balance study was done up to the year 1988, and for the following years it is assumed that the aquifer is subjected to insignificant recharge. With the continuation of pumpage at the level of 1981 (Q = 102.3 m^3/d), it was observed that the interface depth at the well section is always more than the critical depth. Figure 9 graphically represents the depth of the interface beneath the well resulting from an abstraction at a rate of 1.5Q continuously from 1981, and of 2Q starting from 1981 and ceasing at the end of 1989. The analysis indicates that the abstraction at a rate of 1.5Q will not result in any detrimental effect until 1994 and can be expected to be safe in the future. Once the abstraction ceased at the end of 1989 (the later case), a sudden growth of 5.5 m was observed. It is to be noted that this analysis was based on the assumption of no recharge event following the year 1988. This preliminary assessment of long-term withdrawal indicates that the abstraction at 1.5Q is safe for years ahead, whereas abstracting at twice the rate in 1981 results in a critical state.

4. Conclusion

A water balance study incorporating the hydrological budget is conducted to assess the long-term natural recharge to the groundwater system. With geohydrologic information and historical data on aquifer utilization and response, these estimates are integrated with the analysis of dynamic response of the system to assess the long-term sustainable withdrawal rate. This sequence of analysis procedure is illustrated by two case studies. Based on water balance computation, the long-term average annual recharge to Kathmandu Valley was estimated as 45690 m^3/d which is 3 percent of the annual rainfall. The long-term sustainable yield was then estimated based on the dynamic response simulation of the groundwater basin using a two-dimensional numerical model limiting the drawdown to a specified value. The long-term average based on this numerical model study is found to be 45478 m^3/d.

For the coastal island of Mannar in Sri Lanka, a series of monthly recharge sequence over a number of year were estimated based on hydrologic budget and soil moisture balance. The most appropriate recharge sequence is then selected through simulation of movement of fresh-/ salt-water interface under pumpage and recharge condition and comparing the profile with the field observed profile. Following this operational runs were made to predict the long-term withdrawal rate by limiting the upconing effect. The critical upconed height is taken as one-third of the distance between the natural interface location and the bottom of the well. With this condition, one and half times the existing pumping rate was found to be safe for five consecutive year starting from 1988 with the assumption of no recharge sequence prevailing during the period.

References

Amaraweera, H.B.M.P. (1990). Estimation of Safe Yield in Mannar Island, Sri Lanka, M.Eng.Thesis, Asian Institute of Technology, Bangkok, Thailand.

Binnie and Partners (1988). Water Supply for Kathmandu-Lalitpur from Outside the Valley, Prefeasibility Study, Final Report, Water Supply and Sewerage Corporation, Nepal.

Chow, V.T. (1964). Handbook of Applied Hydrology, McGraw-Hill Book Company.

Dagan, G. and Bear, J. (1968). 'Solving the problem of local interface upconing in a coastal aquifer by the method of small perturbations', Journal of Hydraulics Research, 6(1):15-44.

Das Gupta, A. and Amaraweera, H.B.M.P. (1993) 'Assessment of long-term withdrawal rate for coastal aquifer', Ground Water, 31(2):250-259.

Das Gupta, A. and Gaikwad, V.P. (1987). 'Interface upconing due to a horizontal well in unconfined aquifer', Ground Water, 25(4):466-474.

Doorenbos, J. and Pruitt, W.O. (1975). Crop Water Requirement, Irrigation and Drainage Paper 24, Food and Agricultural Organization, Rome.

Hantush, M.S. (1968). 'Unsteady movement of fresh water in thick unconfined saline aquifers', Bulletin, Int'l Assoc. Sci. Hydrol., 13(2):40-60.

McWhorter, D.B. and Sunada, D.K. (1985). Groundwater Hydrology and Hydraulics, Water Resources Publication, Fort Collins, Colorado.

Rao, K.V.R.; Mendis, D.P.J.; de Silva, S.H.G.T. and Perera, H.A.S.A. (1984). Groundwater Investigation for Community Water Supply Source in Mannar Island, National Water Supply and Drainage Board, Sri Lanka.

Rushton, K.R. and Ward, C. (1979). 'The estimation of groundwater recharge', Journal of Hydrology, 41:345-361.

Shrestha, M.N. (1990). Safe Yield of Groundwater Basin in Kathmandu Valley, M.Eng.Thesis, Asian Institute of Technology, Bangkok, Thailand.

Viessman, W. Jr.; Knapp, J.W.; Lewis, G.L. and Harbaugh, T.E. (1977). Introduction of Hydrology, Harper and Row Publishers, London.

THE NUBIAN SANDSTONE BASIN IN NORTH AFRICA, A SOURCE OF IRRIGATION WATER FOR DESERT OASES

Mamdouh Shahin
International Institute for Infrastructural,
Hydraulic and Environmental Engineering
P.O. Box 3015, 2601 DA, Delft, The Netherlands

ABSTRACT. The Nubian sandstone basin in northern Africa is presented in this paper. This basin has total a total surface of about 1.8 million km^2, encompassing the north-western part of the Sudan, Egypt west of the Nile, the extreme north-east of Chad and southern and eastern Libya. This surface is characterized by its extreme aridity. Groundwater is present in the Nubian basin in a range of geological strata, and has been used since time immemorial for small scale potable supplies and for limited irrigation purposes. The only inhabited areas are located in a series of depressions called the oases, the most important of which are the Kharga, Dakla, Farafra, Bahariya and Siwa in Egypt, and Kufra in Libya. The last few decades have witnessed accumulating evidence of the huge quantities of fresh groundwater the Nubian sandstone contain in storage and can transmit. Safely exploited aquifers can secure the success of carefully planned agricultural development projects. Thirty-five years ago the Egyptian administration supported desert reclamation and development by initiating the New Valley (principally Kharga and Dakhla Oases) project. A few years later the Libyan authorities began the work in the Great man-made river project, to which the Nubian basin underlying the Kufra Oasis is a source of supply. Both projects aim at expanding the irrigated surface in the relevant oases. The present paper describes the hydrogeology of the Nubian basin. It also critically reviews the exploitation of its water for land irrigation and reclamation purposes.

1. HISTORICAL INTRODUCTION

By the time of the IVth Dynasty in Egypt, the Old Egyptians became skilful at sinking shafts through soft rocks like limestone and sandstone. Two wells were sunk in the sandstone on the road to the quarries in the Nubian Desert at distances of about 63 and 77 km from the Nile. The estimated depth is over 70 m. Well-sinking was also practised during the XIth Dynasty (2100-2000 B.C.) on the road to other quarries where 12 wells are claimed to be dug. Murray (1955) adds that the well which is still existing there is 32 meters (100 ft) and is of great antiquity.

There are further evidences that King Seti and his successor Ramsis II,

V. P. Singh and B. Kumar (eds.), Subsurface-Water Hydrology, 193–210.

and later Ramsis IV, all in the period from 1310 to 1160 B.C., had dug several deep wells on the roads from the Nile in the southern part of Upper Egypt to the locations of the gold mines in the Eastern Desert. Additionally, Murray (1953) mentioned that the earliest inscription indicating to the installation of a water well in an oasis (Dakhla Oasis) in the Western Desert dates back to about 905 B.C. There, the thickness of the clay deposits before reaching the groundwater aquifer varies from 32 m (100 feet) to 100 m (300 feet).

Shortly after 512 B.C. Admiral Scylox was sent by Darius I, King of Persia, to El-Kharga Oasis in Egypt (see the map, Figure 1). According to Butler (1933) Scylox introduced the Persian method of irrigation by means of underground conduits (qañat) fed by water from the sandstone strata where it used to be collected in faults.

The development of irrigation in El-Kharga Oasis went fairly well during the Ptolemaic period (323-30 B.C.). Artesian water was fully exploited during the Roman Empire (30 B.C.-A.D. 395) to meet with the requirements of cultivation. Thompson and Gardner (1932) reported that mining of groundwater became aggravated during the Byzantine period (A.D. 395-638) with the establishment of the Christian settlements near the oases in the Western Desert. Later those remarkable works were left to deteriorate. However, the priesters in the isolated monasteries and convents in the deserts remained to maintain the wells in good condition as they were the only available resources of water there.

The groundwater situation underneath the Libyan Desert in general and the oases in particular began to attract the attention of many an investigator since the beginning of the twentieth century. To be mentioned specifically are the important studies of Ball (1927) from which he was able to draw the first map of the contour lines of equal groundwater levels covering the Western Desert in Egypt and the north-western corner of the Sudan. This map was later modified by Hellström (1940), who also completed it by plotting the flow lines, i.e. he produced a complete flownet. One of the important conclusions drawn by Hellström was to locate the main intake beds of the groundwater underlying the Desert, which he claimed to be east of Erdi and Ennedi highlands (northeast of Chad) and north of Wadi Hawar (northwest of the Sudan). The findings of Ball and Hellström were reviewed and updated by Murray (1952). He emphasized that the groundwater underlying the oases is not connected to the Nile water, instead it is fossile water.

CLIMATE

The climate is generally characterized by high temperature variations. In the summer an air temperature of 45-50°C by day is not exceptional. The night (or early morning) temperature can easily fall to 0-5°C. At Kufra in Libya the January minimum is 4.0°C, while the July maximum is 45.3°C. The mean relative humidity of the air outside the heavily irrigated oasis is in the order of 20%. In Egypt the mean January and July temperatures at the Kharga and Dakhla oases are about 13°C and 31°C, respectively, with a mean daily range of variation of about 17°C. The mean annual relative humidity of the air is about 40% at the Dakla Oasis and 46% at the kharga Oasis. The cloudiness varies from almost zero in the summer time to 1.2 oktas in the winter time.

There is hardly any rainfall on the Egyptian desert at present. In Libya the rainfall inland far from the Mediterranean coast is 10 to 20 mm yr^{-1}. Rare but occasionally high intensity (torrential) rainstorms may, however, produce such values in one single hour. These extreme rainfall events may cause rare floods. The North African region has been subject to aridifying effects for the last 15000 years or so, although during that period some pluvial interruptions have occurred.

The high figures of the short wave radiation combined with the aforementioned climatic data lead to high rates of potential evapotranspiration and evaporation. These rates are usually in the order of 2500 to 3000 mm. The evaporation occurs in depressions, where groundwater is present at shallow depth, the so-called sabkhas. Although the top soil is dried by atmospheric conditions, the underlying soil is constantly wet by the capillary ascent from the shallow water table. As a consequence, liquid flow is dominant in the deeper soils, while as vapour transfer is dominant in the upper soils. The interface between the two movements is termed evaporation front. The annual evaporation from the area around the Qattara depression has been estimated from numerical models and satellite measurements at 2000 to 2700 million m^3 (Menenti et al, 1991).

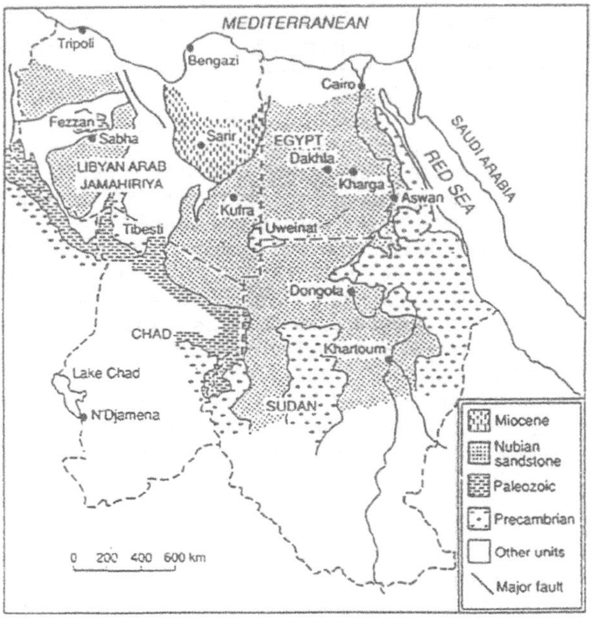

Figure 1 - Regional extent of the Nubian sandstone aquifer.

GEOLOGY

The area under discussion is dominated by lower Cretaceous and later sediments. The lower Cretaceous continental sandstones are often referred to as Nubian sandstone. This name derives from the outcrops in Nubia in the north of the Sudan and the south of Egypt. The Nubian facies have been protected by the rigid substratum of the basement complex, which were fused and carbonized into one immobile mass in the Precambrian. They derive from the Precambrian and from reworked sandy Palaeozoic deposits and have been altered by metamorphic processes. During the process of peneplanation they have gradually been deposited on the periphery of the culminating outcropping regions of the old basement complex.

In addition to the geological setting of the region shown in the map, Figure 2, two geological sections are shown in Figures 3a and 3b. The first, after Llyod (1990), extends from southern to northern Egypt, and the second, after Wright et al (1982), through Kufra and Sirte Basin in Libya.

The Nubian formation in the Kharga Oasis, for example, consists mainly of thick beds of sandstones intercalated with thin layers of shales and clays. The sandy members of this formation constitute the groundwater aquifer with a net sand thickness of about 400 m. The clay and mudstone members exposed at the surface furnish most of the agricultural soils of the Oases. The depth to the basement rock varies in the Kharga Oasis between 600 m and 800 m.

South of the parallel of Tazerbo (see map, Figure 2), the Kufra basin is well delineated on the periphery by Palaeozoic outcrops mostly made of continental sandstones. The central part of the basin is occupied by Mesozoic continental sandstone outcrops occasionally covered by sand dunes or alluvial deposits. North of Tazerbo, a Cretaceous and lower Tertiary marine transgression overlays the Mesozoic and Palaeozoic sandstones, the thickness of which decreases rapidly to the north. After the Eocene, a continental environment prevailed. Pot-Eocene deposits consist chiefly of sand, sandstone and clay with some limestone. Their total thickness reaches 800 to 900 m in the centre of the Sarir well field where they form another freshwater reservoir. Further north the Pot-Eocene continental deposits pass gradually to shallow marine carbonaceous formations with many evaporitic intercalations characteristic lagoonal and estuarine conditions. This change of facies corresponds also to a change of the water quality which becomes brackish and even salty as soon as groundwater flows through evaporitic deposits (Pallas, 1979).

The Palaeozoic sediments were deposited on the basement rocks some 300 millions years before present. These deposits consist mainly of sand and permeable sandstone, which help store and transmit the groundwater. Further deposition of sand and Nubian sandstones took place in the Mesozoic era about 120 million years ago. These formations extend in the northern half of the Sudan, except in the middle, and together with most of the surface of the Western Desert in Egypt, the south-eastern part of Libya and northern Chad form the extensive Nubian basin in northern Africa. The thickness of the basin in the Sudan is estimated at 3000 m.

Hydrogeology

Nubian Sandstone in the Western Desert of Egypt

The water levels in the various wells scattered in the oases far south of the coastal strip suggest that they belong to one huge body of ground

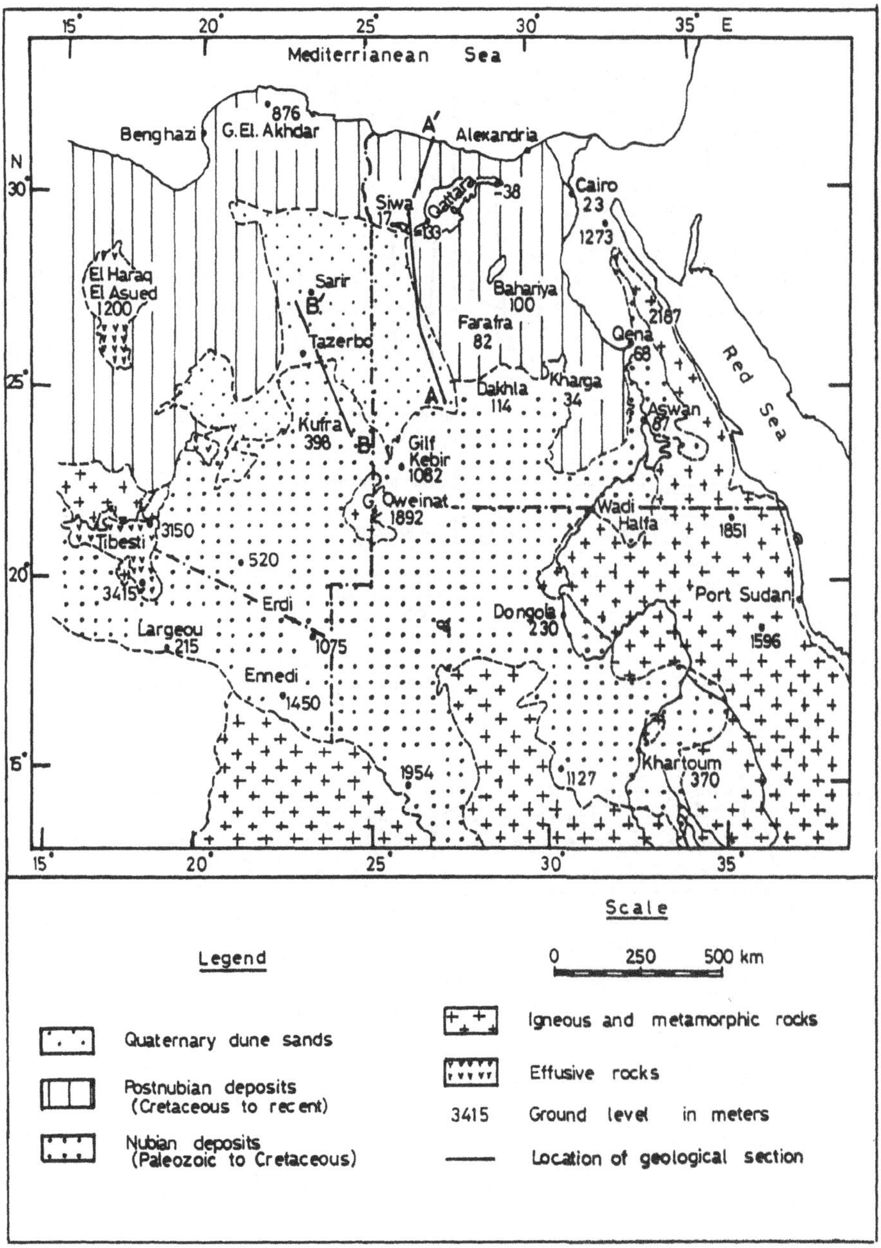

Figure 2. Regional geologic seting of the Eastern Sahara.

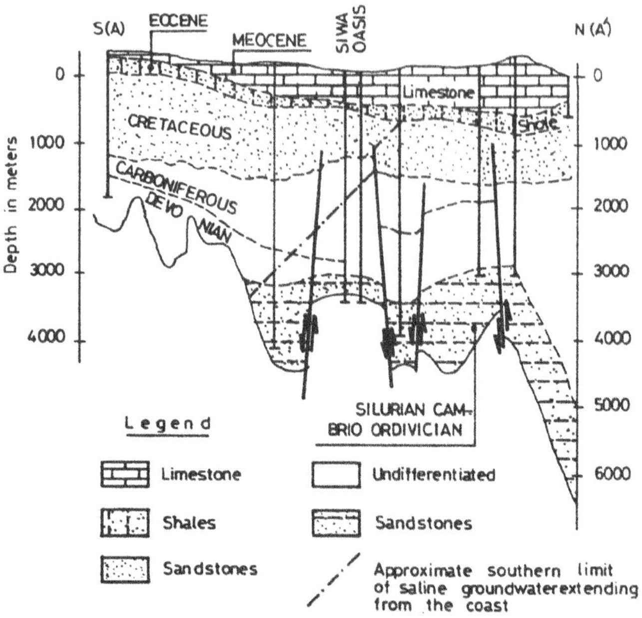

Figure 3a. Geological section A A' in the western Desert, Egypt

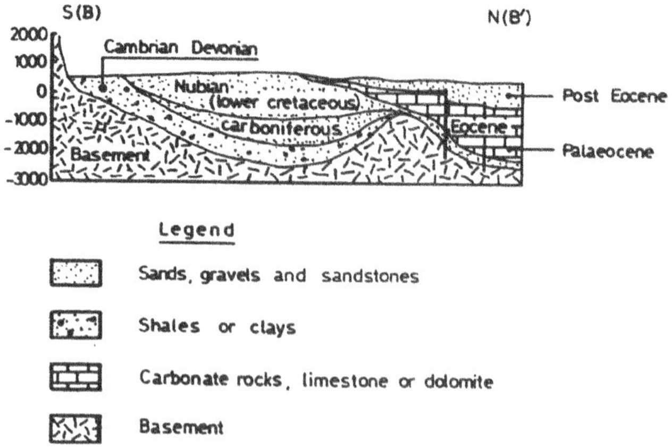

Figure 3b. Geological section B B' in the Nubian Basin, Libya.

water. Around the beginning of the century G. Rohlfs developed the idea, which has not been changed since then, that all oases of the Libyan Desert derive their groundwater supplies from the rainfall in Western Sudan. Referring to Soliman (1954), Sir Lyons recommended in 1908 an investigation to verify whether a general water-table does exist underneath the Libyan Desert. John Ball (1927) presented for the first time a map showing contour lines of equal piezometric head. This map is shown in Figure 4a. In 1935, Sanford K. extended those contour lines to cover the northwest corner of Sudan.

According to Ball, the origin and source of artesian water supply to the desert oases is a vast subterranean flow of water coming from the Erdi and Ennedi region on the border between the basin of lake Chad and the basin of the Nile River. This underground water flows in the directions of northwest, north and northeast thus feeding, among others, El-Kharga, El-Dakhla, Farafra and Siwa Oases. He also observed that the Nile between the first and second cataracts cuts its channel through water bearing Nubian Sandstone strata, through which flows the groundwater feeding the oases.

B. Hellström (1940) studied the findings of Ball in the light of more recent data. That study enabled him to draw the flownet presented in Figure 4b, and thereupon to conclude that the main intake beds are located east of the Erdi and Ennedi highlands (northeastern corner of Chad) and north of Wadi Hawar (northwestern part of Sudan). Among his findings is that Quatarra depression, the oases southwest thereof and the Fayoum depression intercept some of the flowing groundwater. The amount intercepted was estimated at 3.1×10^6 m^3 d^{-1}. Hellström used this estimate in computing the permeability coefficient of the Nubian Sandstone. This he found to fall in the range of from 2 to 20 m d^{-1}. Note that the range adopted recently in some model studies is from 2 to 10 m d^{-1} . The same reference estimated the average gradient of the piezometric pressure from the intake bed down to the Quattara Depression as in 1 in 2000. This figure leads to an approximate value of 15 m yr^{-1} for the average flow velocity.

In 1952, G.W. Murray presented a report in which he discussed the works of both Ball and Hellström after considering more recent data and correcting some of the old ones. The new contour lines of the equipiezometric head are shown in Figure 4c.

A more recent flownet diagram has been developed by Ezzat in 1974. This is in fact an updated version of the previous one, developed by Murray. The former is illustrated by Figure 4d. This map too emphasizes that the groundwater moves from the intake bed in northwestern, northern, northeastern and southwestern directions. The flow gradient is in the range 1/1500 to 1/2500. Ezzat (1976) estimated the total annual recharge to the Western Desert Sandstone aquifer at about 3.58 x 106 m^3 d^{-1}. Of this amount 1.82×10^6 m^3 d^{-1} are passing through the Libyan front, 0.52×10^6 m^3 d^{-1} passing the Gilf el-Kebir uplift (southwestern corner of Egypt), and 1.24×10^6 m^3 d^{-1} are passing through the Sudanese border. The various discharges and rates of exploitable water per oasis are given in Table 1.

Hammad (1969) reported that there is a guaranteed steady daily recharge of 174×10^3 m^3 and 384×10^3 $m^3$to the Kharga and Dakhla Oases, respectively. Some other investigators do not strongly support these figures. The same author concludes also that the global average figures for the aquifer transmissivity are between 1000 and 1800 m^2d^{-1}. The corresponding storativity varies from 2×10^{-4} to 4×10^{-4}.

In order to check the behaviour of the Nubian Sandstone aquifer under different climatic conditions, a model study has been performed and reported by Chow and Wilson (1981). The model, known as AQUIFEM, represented a section extending from the Ennedi highlands of northwestern Sudan to the

Nile Delta, Egypt, passing through the Farafra and Bahariya Oases. The geological formation in the northwestern Sudan was modelled as having a constant thickness. In Egypt, the basement dips from the southwest to the northeast along the chosen section line. The thickness of sand increases from 500 m near Kufra Oasis in Libya to 2 100 m near Farafra and Bahariya

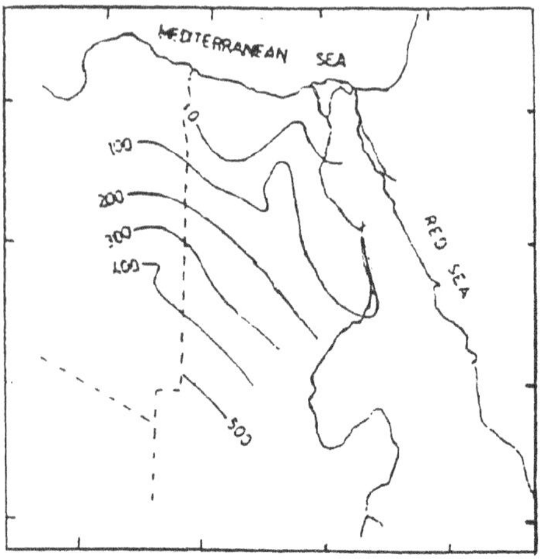

Figure 4a - Equipotential lines of groundwater in the W. Desert, Ball (1927).

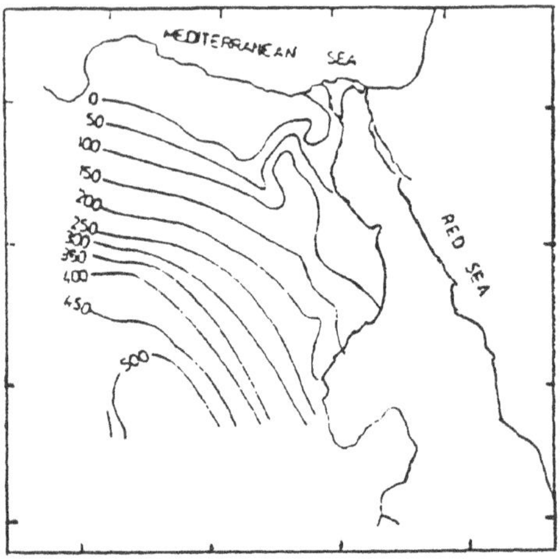

Figure 4c -Equipotential lines of groundwater in the W. Desert, Murray (1952).

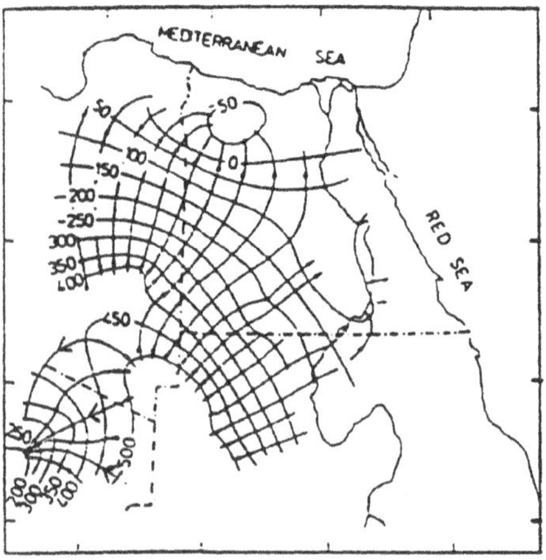

Figure 4b - Flow net of groundwater in the W. Desert, Hellström (1940).

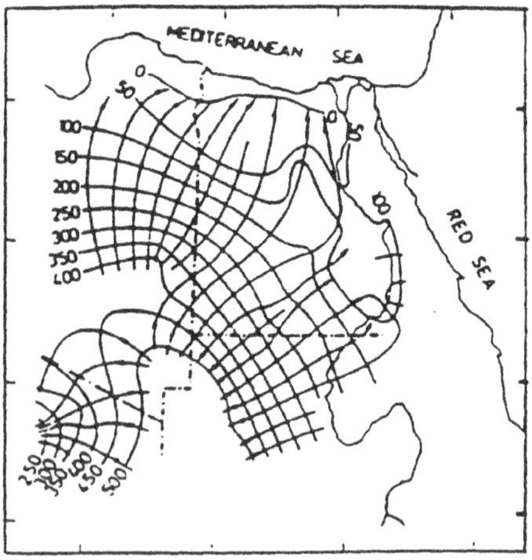

Figure 4d - Flow net of groundwater in the W. Desert, Ezzat (1974).

Oases, and thins to about 900 m at Wadi el Natrun close to the Nile Delta (see Figure 5). The coefficient of permeability adopted in the model was 6 m d^{-1}, the storativity 1 x 10^{-5} and the specific yield almost 0.10.

Three major mechanisms have been hypothesized and each of them tested. In the first hypothesis the precipitation in the southern regions (intake beds) was considered to be sufficient to maintain the piezometric gradients as observed currently. As such, a recharge of 10^{-5} m yr^{-1} was applied over 200 km cross-section in the Ennedi highlands area. The results obtained from model simulation run are shown in Figure 6a. This result suggests that the recharge applied is sufficient to maintain the piezometric surface near its current levels in the steady state. Since the recharge is much less than the precipitation, the latter must be more than adequate to maintain the piezometric levels.

Table 1 - Discharge data of the Western Desert Oases

Area	Amount of water, m^3 d^{-1}, corresponding to			
	Discharge wells	Natural losses	Total	Exploitable water
Siwa Oasis	120 000	300 000	420 000	384 000
Quatarra Depression	--			--
Farafra and Bahariya		1400 000	1400 000	
Oases	145 000			546 000
Dakhla Oasis	537 000	400 000	545 000	1395 000
Kharga Oasis	225 260	141 000	698 000	427 000
South Kharga	--	190 260	416 000	--
		100 000	100 000	

The second hypothesis is that the recharge from past periods of high precipitation is sufficient to maintain present levels of piezometric head. In this case recharge to the Nubian sandstone was applied until a steady state has been reached. An initial recharge rate of 0.01 m yr^{-1} was applied

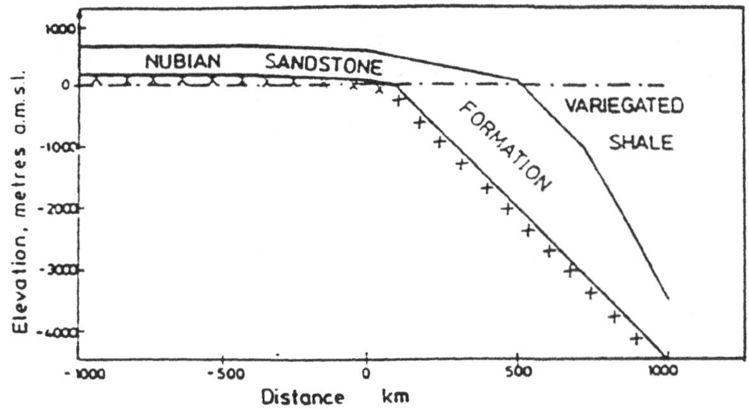

Figure 5 - Geological section adopted in the AQUEFEM model.

to simulate the current piezometric levels after having 6000 years of no recharge. The results obtained are shown graphically in Figure 6b.

The third and last hypothesis is that present-day precipitation is adequate to maintain the piezometric head at its current level. A recharge of 0.001 m yr^{-1} was applied to the outcrop areas only. This was found insufficient to produce piezometric heads at their current levels, although a net recharge of 0.005 m yr^{-1} was able to yield the current piezometric levels (see Figure 6c). This result shows that current precipitation alone cannot be the source of all groundwater.

From the above discussion the conclusion that can be drawn is that the recharge from the southern highlands can easily maintain the piezometric head levels. Recharge from past fluvial periods is a "plausible" mechanism.

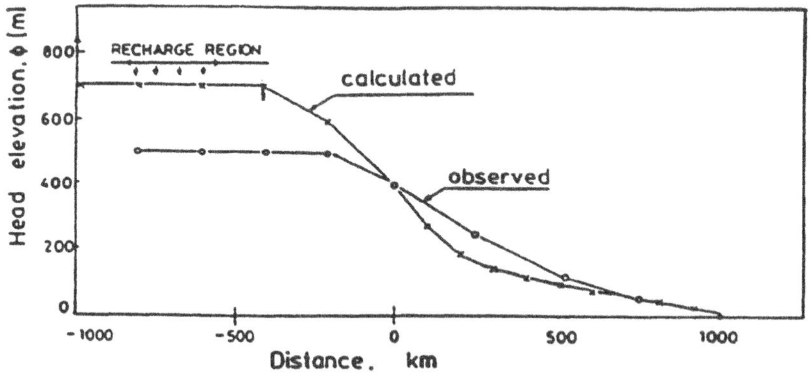

Figure 6a - The calculated heads as obtained from the first hypothesis.

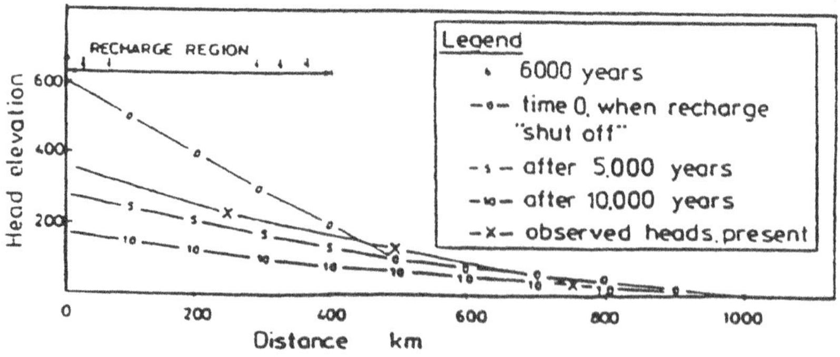

Figure 6b - The calculated heads as obtained from the second hypothesis.

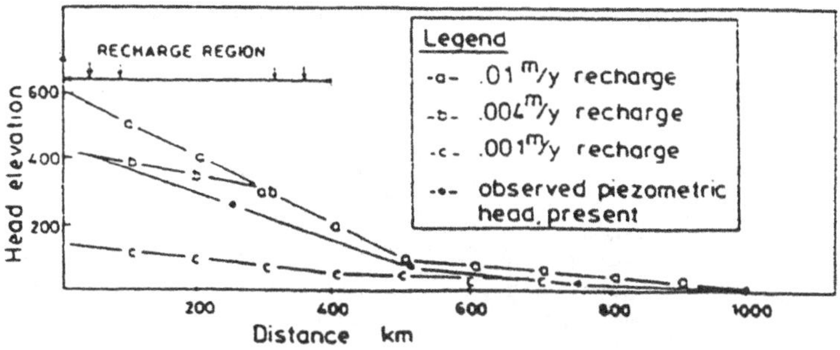

Figure 6c - The calculated heads as obtained from the third hypothesis.

Current precipitation is an unlikely mechanism to maintain groundwater levels by itself. However, it can supplement other mechanisms. It is probable that the groundwater in the Nubian Sandstone is a combination of these mechanisms and of others which may not have been hypothesized. (Chow and Wilson 1981).

The central part of the basin is situated south of the N 26° parallel and occupies a surface of more than 250 000 km². It's maximum depth can easily reach 3 000 m. The thickness of the sandstone layer decreases, however, rapidly to the north. The basin is occupied by Mesozoic continental sandstone outcrops occasionally covered by sand dunes or alluvial deposits. The precipitation is limited at present to just a few million m^3 per year (Pallas, 1979). It is likely that the groundwater flow is caused by the slow emptying of the groundwater reservoir underlying the Sahara, which was filled up by the rainfall during the pluvial episodes, and the inflow from the Sudan and Chad. The natural discharge consists of the outflow reaching the north (Sarir basin) and the north-east (Egypt). To this outflow should be added the water evaporated from the lakes and sabkhas. Lloyd (1990) reported two hydrogeological features of the Nubian system at Kufra; firstly, although the Kufra development project is within the regional (in confined areas, the local geology is layered such that a leaky condition in part dictates the deep well yields, secondly, the effective radius of the cone of depression is only 25 km after 24 years of abstraction of some 90 - 120 10^6 m^3 yr^{-1}, indicating a considerably higher regional specific yield value than the 0.15 figure adopted in earlier investigations. Long-term observations have shown that the drawdown caused by abstraction of a maximum of 120 x 10^6 m^3 has resulted in a drawdown of 0.2 m some 30 km far from the well field. The thickness of the Nubian aquifer is taken for design purposes as 1000 m with a storage coefficient of 1.1 x 10^{-4}-1.5 x 10^{-2}, transmissivity of 300-3 500 m^2 d^{-1} and hydraulic conductivity of the confining layer of 0.002 m d^{-1}. The groundwater heads and their distribution within the Kufra basin were plotted by Pallas (1979) when the abstraction from the Kufra well yield had been 90 x 10^6 m^3 in the year 1977-1978. Part of that map is redrawn in Figure 7. A more recent map has been prepared by Wright et al (1982) and later adjusted by Lloyd (1990) after the abstraction has increased to 120 x 10^6 m^3 in the year 1987-1988.

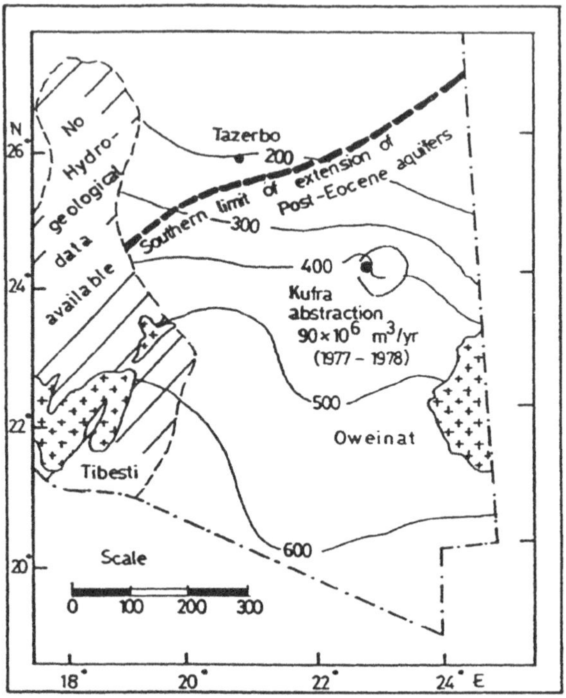

Figure 7 - Groundwater heads in the Nubian sandstone in Libya.

The Nubian Sandstone System in the Sudan

The information so far available shows that the Nubian sandstone system covers a surface of about 660 x 10^3 km^2 in the Sudan. This area comprises a number of basins with an estimated total storage capacity of about 16.3 x 10^9 m^3 and annual recharge between about 42 to 257 x 10^6 m^3 y^{-1} as can be seen from Table 2.

Hedayat (1986) presented a map showing the groundwater heads in the Nubian sandstone in the Sudan. Figure 8 shows this map after some adjustment of the contour lines in the northwestern corner where the boundaries between the Sudan, Chad, Libya and Egypt meet.

To further investigate the origin of the groundwater in the Nubian aquifer system the research project "Geoscientific Problems in Arid Areas was carried out and the field-work in Egypt was completed in the period 1984-1987. The investigation was accomplished by means of numerical models of flow and transport. Groundwater balances were established for different subregions for simulating of steady and transient-state conditions.

The principal conclusions drawn from that investigation (Brinkman et al, 1987) can be summarized as follows:

Table 2 - Storage and recharge data of the Nubian basin in the Sudan

Groundwater basin	Storage capacity, 10^6 m^3	Recharge, 10^6 m^3/yr upper limit	lower limit
Nubian sandstone	5500	136	3.7
Sahara Nilotic	9740	20.6	20.6
Central Darfur	794	47.6	12.8
Al-Nehoud	136	15.4	1.5
Sáng Annáam	134	14.8	1.3
Atbara mouth		23.0	2.7
Total	16304	257.4	42.6

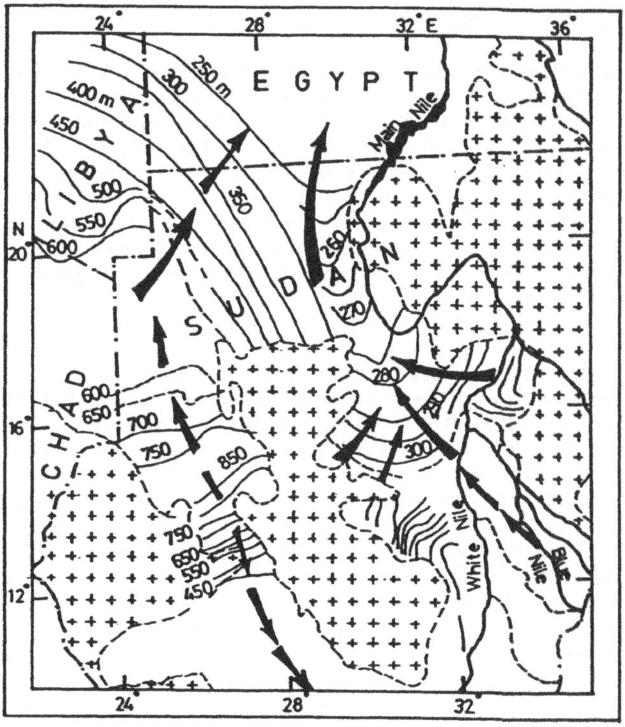

Figure 8 - Groundwater heads in the Nubian Sandstone in the Sudan.

- Groundwater was recharged over large areas in the whole Nubian sandstone system. During those recharge periods no large-scale flow occurred. Infiltration supporting equilibrium conditions stopped some 8 000 years ago. Since then the natural groundwater situation, i.e. before artificial extraction began, has been in an outflow process.

- Paleo recharge also occurred in the northern confined part of the aquifer. The old age of groundwater in the intermediate layers can be explained by the low flow velocity.
- The water balance for steady and transient states show a deficit in the confined part of the aquifer. This deficit is made up for by the regional flow from south to north.
- It appears that the greatest part of the Nubian aquifer system has been recharged over millions of years, and that the aquifer probably has never been depleted. Changes in groundwater level occurred only in the upper tens or a few hundred metres. Even during very extensive arid periods, the aquifer could have not been depleted to a level lower than the lowest discharge areas.

Abstraction of Groundwater for Desert Irrigation

About the end of the nineteen forties the General Inspectorate for Desert Irrigation was established. This was replaced in the early sixties by the Egyptian General Desert Development Organization (EGDDO) and a few years later by the Egyptian General Desert Development Authority (EGDDA). This organization (later authority) kept itself busy with groundwater in the Desert Oases; abstraction for irrigation development projects, deep well drilling, water distribution and irrigation systems, land reclamation, etc. The Desert Institute of Egypt, which is a much older establishment, is engaging itself essentially in research work, in several parts of the desert and not in the oases only.

Upon the request of the former Desert Irrigation Department, Paver, G. and Pretorius, D. (1954) prepared a report on the hydrogeological investigations in El-Kharga Oases. That report showed that the total discharge of the shallow wells in El-Kharga Oasis fell from 123 200 m^3 d^{-1} (354 wells) in 1941 to 105 800 m^3 d^{-1} (412 wells) in 1953. Contrarily, the yield and number of wells in El-Dakhla Oasis increased from 109 600 m^3 d^{-1} (577 wells) in 1900 to 222 480 m^3 d^{-1} (872 wells) in 1925 to 229 280 m^3 d^{-1} (949 wells) in 1931 and 253 500 m^3 d^{-1} (805 wells) in 1953. the report of Paver and Pretorius (1954) included an estimate of the yield and the number of deep wells needed to meet with the requirement for reclaiming and for irrigating new land. The conclusions contained in that report as well as in other reports were considered while laying out the plans for implementing the project known as the "New Valley". The project aimed at expanding the agricultural land in and around the oases by irrigating new lands from free-flowing groundwater. To fulfil this objective new deep wells (hundreds of metres in depth) have been drilled and many of the old wells have been cleaned and brought into function. The immediate result is that the abstraction from the deep aquifer underlying the Kharga and Dakhla Oases increased steadily from about 20 x 10^6 m^3 by 1956 to about 55 x 10^6 m^3 in 1960 and further to about 210 x 10^6 m^3 by 1968. The obvious consequence is the rapid fall (30 m) of the groundwater head in the period from 1962 upto 1969 (Margat and Saad, 1984). This period was followed by a narrow fluctuation in the volume of groundwater annually abstracted so that from

1970 upto 1975 the further drop in head did not exceed 3 m (see Figure 9).

The good quality of the groundwater in the deep aquifer and the artesian flow condition were among the factors that encouraged the rather heavy exploitation of the basin in the period 1960-1967.

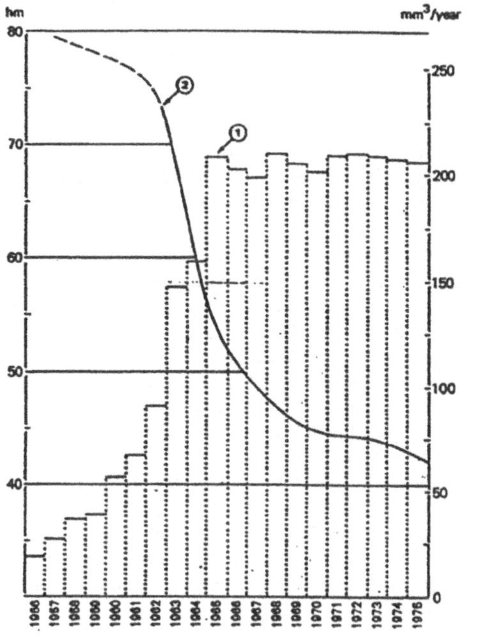

Figure 9 - Change of abstraction rate and decline of groundwater head in the artesian aquifer underlying the Kharga and Dakhla Oases in Egypt, 1956-1975.

Freshwater occupies the surface from the southern boundary of the artesian basin up to the parallel line 28° north. This water is essentially of meteoric origin with a total salinity from 100 to 800 ppm. The groundwater contained in the uppermost part of the aquifer, however, is generally less fresh than the water in the lower part.

The sharp decline followed by the gradual decline in the groundwater heads in the deep aquifer led to a considerable reduction in the flow of some wells, some others even ceased to flow naturally at all. By the end of the 1970's it became clear that land irrigation in the oases cannot be maintained without pumping, else the irrigated surface has to be reduced to just what the natural flow can supply. The development in the situation of the Kharga Oasis may help illustrate the consequence of over exploitation of the aquifer. In 1962 the annual abstraction on the basis of 100% free flowing wells was about 119 x 10^3 m^3 d^{-1} or 43.5 x 10^6 m^3 yr^{-1}. This abstraction grew to almost 79 x 10^6 m^3 in the year 1980 on the basis of 43% (corresponding to 33.8 x 10^6 m^3 yr^{-1}) free flow and the rest is pumped flow.

Future plans are laid down on the basis of total annual abstraction of 186×10^6 m^3 yr^{-1} of which 156×10^6 m^3 go for irrigation and the remaining 30×10^6 m^3 yr^{-1} will be consume in phosphate mining-processing . The projected drawdown will be in the order of 90 m. Since the desert conditions make the annual irrigation requirements as high as 22 000 m^3, the 156×10^6 m^3 will help irrigate only 7000 ha. The ever increasing pumping costs are gradually lessening the economic viability of the existing as well as projected irrigation schemes.

In his paper on the future of groundwater in African Sahara Desert, Hammad (1969) reported that the free flow to the Kharga Oasis is limited to 174 x 10 m^3 d^{-1} and in the Dakhla Oasis to 384×10^3 m^3 d^{-1}, i.e. the two together to just 204×10^6 m^3 yr^{-1}. Additionally, he advised not to use pumps, else the cost of land irrigation will become prohibitive. A similar recommendation was given later (1987) by Brinkmann et al. "While the aquifer system can be kept filled by just a few millimeters recharge each year, some millimeters (5 to 8 corresponding to 18 000-30 000 m^3/ha yr^{-1}) are needed each day for the purpose of irrigation. So, if the latter has to be supplied for irrigating large surfaces of the desert, the groundwater level would drop then by 15 to 20 m yr^{-1}. As such, the only possible exploitation is the irrigation of small areas of selected locations. The groundwater in the surrounding areas that was formed during the last thousands of years has been exhausted due to its movement towards the centers of the cones of depression.

The pressing need for more acres to be brought under irrigation has led the concerned desert development authority not to follow the advice and proceed with pumping water for irrigation.

The groundwater in the deep aquifer underlying the Kufra Oasis, Libya is of good chemical quality. The total dissolved solids are in the range 150 to 350 ppm. The shallow aquifers may have a much higher salinity. This situation encouraged the Libyan authorities to intensify. The exploitation of the Kufra deep aquifer for the purposes of irrigation and domestic water supply.

The Kufra production project (KPP) well field consisted of 102 production wells, whereas the Kufra settlement project (KSP) consisted of 35 wells. The depth of the wells in the KPP field is 350 m and in the KSP well field is 475 m. These two well fields were completed in 1968 and 1974, respectively. The total abstraction in the period 1960-1987 has slightly exceeded 1.5×10^9 m^3 (see Figure 10). The KPP abstracts annually about 90×10^6 m^3 and the KSP wells abstract 30×10^6 m^3, i.e. the two well fields together abstract 120×10^6 m^3 yr^{-1} annually, as from 1987. Both draw their water from the deep aquifers. These figures are supposed to reach 120 and 96.5×10^6 m^3 yr^{-1} for the KPP and the KSP respectively at the end of the final phase of the project. The aim is that the KPP helps irrigate 10 000 ha and KSP 5 500 ha, i.e. a total surface of 15 500 ha in the Kufra Oasis consuming about 220×10^6 m^3 each year. The past twenty years have witnessed a decline of more than 40 m in the groundwater head. With abstraction of 220×10^6 m^3 yr^{-1} the decline in the project area of the water level will range from 50 to 130 m after 50 years of operation. As a consequence the pumping level in the wells, measured from the ground surface, will be in the range from 70 to 150 m.

Brinkmann et al (1987) concluded that desert irrigation schemes in Egypt and Libya form extraction cones. The groundwater in the surroundings has been exhausted due to its movement towards the centers of these cones.

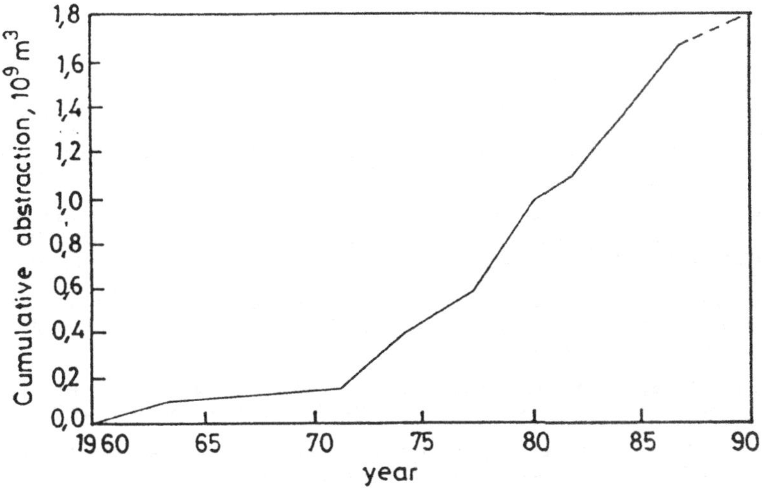

Figure 10 - Abstraction of groundwater from the Kufra basin, Libya, in the period 1960-1990.

CONCLUSIONS

From the above overview one can fairly conclude that:

i - Groundwater abstraction is unavoidable to sustain life in the desert oases.

ii - The quality of groundwater is good and suitable for domestic as well as irrigation purposes. The latter constitutes no less than 90% of the abstracted volumes.

iii - The annual recharge to the Nubian sandstone aquifer is quite limited. Any abstraction in excess of the natural recharge leads to the mining of the basin, a situation which should not be allowed to develop unless economically and/or socio-economically justified.

iv - The crops so far raised in the irrigated areas are of the traditional and not the market cash type. Moreover, it is claimed that the wasted water is on the high side, bringing the efficiency of water use to a rather low level.

v - Deep well drilling has definitely helped in collecting data about geology and hydrogeology of water and oil bearing strata. This has been paralleled by the continuous development in groundwater modeling. Field data and models combined together have improved the technical knowledge with regard to the abstraction of groundwater from the reservoirs in the desert. For one reason or another the economical and/or social-economical aspects are hardly known. This state of affairs renders any decision making based on cost-benefit analysis or any similar objective function quite difficult.

REFERENCES

Balls, J., 1927, 'Problems of the Libyan Desert', Geogr. J. **70** (1), (2), and (3).

Brinkmann, P.J. Heinl, M., Hollander, R. and Reich, G., 1987, 'Retrospective simulation of groundwater flow and transport in the Nubian aquifer System', Berl. Geowiss. Abb., Reihe A/Band **75.2**: 465-516, Dietrich Reimer (Publ.), Berlin.

Butler, M.A., 1933, 'Irrigation in Persia by Quanats: an ancient method of collecting and conducting water in long underground galleries in use today'. Civ. Engr. ASCE **3**: 69-73.

Chow, J.S. and Wilson, J.L., 1981, 'A quantitative review of Nubian Sandstone regional aquifer behaviour', Proc. Int. Conf. on Wat. Resour. in Egypt: 363-382, Cairo, Egypt.

Ezzat, M.A., 1976, 'Regional groundwater model; the New Valley project', Working document 2, UNDP/FAO report, AGON:EGY **71/561**, Cairo, Egypt.

Hammad, Y.H., 1969, 'Future of groundwater in African Sahara Desert', Proc. Am. Soc. Civ. Engrs., J. Irr. & Dr. Div. **95**(4): 563-580.

Hedayat, A.A., 1986, 'Water resources in the Republic of Sudan, and their utilization', report (Arabic text) submitted to the panel on Water Resources in the Arab Region and their Utilizations, Kuwait.

Hellström, B., 1940, 'The subterranean water in the Libyan Desert', Geografiska Annaler **34**: 206-239, Stockholm.

Lloyd, J.W., 1990, 'Groundwater resources development in the Eastern Sahara', J. of Hydrol. **199**: 71-87.

Margat, J., and Saad, K.F., 1984, 'Deep-lying aquifers: water mines under the desert?' Nature and Resources, **XX**(2): 7-13, UNESCO, Paris.

Menenti, M., Bastiaanssen, W.G.M., Hefny, K. and Abd el Karim M.H., 1991, 'Mapping of groundwater losses by evaporation in the Western Desert of Egypt', Report series **43**: 116, Wageningen/Cairo.

Murray, G.W., 1952, 'The artesian water of Egypt', Pap. **5** Survey Department of Egypt, Cairo.

————————, 1953, 'Water from the desert: some ancient Egyptian achievements', Geogr. J. **121**: 171-181.

Pallas, P. 1979, 'Water resources of the Socialist People's Libyan Arab Jamahiriya', Report submitted to FAO, Tripoli.

Paver, G.L. and Pretorious, D.A., 1954, 'Hydrogeological investigations in Kharga and Dakhla Oases. Tech. Rep. **4**, Inst. du Désert d'Egypte, Cairo.

Soliman, H., 1954, 'Groundwater table in the Libyan Desert,' 2nd ICID Congr. on Irr. & Dr., Trans. **III**: 6.149-6.160.

Thompson, C.G. and Gardner, E.W., 1932, 'The prehistoric geography of Kharga Oasis', Geogr. J. **LXXX** (5): 369-409.

Wright, E.P., Benfield, A.C., Edmunds, W.M. and Kitching, R., 1982, 'Hydrogeology of the Kufra and Sirte Basins, eastern Libya', Q.J. Geol., **15**: 83-103.

DELINEATION OF GROUNDWATER RESOURCES IN DROUGHT PRONE AREA OF UPPER GODDUVANKA RIVER BASIN, CHITTOOR DISTRICT, ANDHRA PRADESH USING ELECTRICAL RESISTIVITY METHOD

T.V.K. REDDY[1], S.S. REDDY[2] and N. JANARDHANA RAJU[3]

1. Reader in Geology, Department of Civil Engineering, S.V.University, Tirupati-517502, INDIA.
2. Lecturer in Geology, Department of Civil Engineering, Pondicherry Engineering College, Pondicherry.
3. Research Associate, Department of Geology, S.V.University, Tirupati-517502, INDIA.

ABSTRACT : Electrical resistivity method has been used to delineate groundwater resources in drought prone area of Upper Godduvanka river basin, Chittoor district, Andhra pradesh. 71 vertical electrical soundings were conducted using Wenner configuration. The data was interpreted by partial curve matching with the help of master curves. Contour map of apparent resistivity for the separation a = 30 m is prepared to delineate high and low resistivity zones. Based on the results, three vertical geoelectrical cross sections are prepared to know the distribution of groundwater resources along the respective profiles. Depth to basement contour map and finally probable groundwater potential zones are delineated for future exploitation. The study indicate that the groundwater resources are mainly confined to the sand and weathered granite zones of the study area.

INTRODUCTION

With the ever increasing demand for water supply and the inadequate surface water especially in the drought prone areas, attention is turned towards groundwater resources. The present study area depends mostly on groundwater resources for the development of the region. The basin comprises of crystalline rocks of Archean age. The major rock type exposed in the study area is granite which is intruded by dolerite dykes. The river course is mostly concealed by alluvium and soil. The identification of structurally favourable zones of groundwater viz., lineaments, faults, fractures and joints have gained importance in the development of groundwater potential zones in hardrock.

Of all the surface geophysical methods, the electrical resistivity method has been applied most widely for groundwater investigations (Todd, 1959). The search for groundwater in hard rock areas can be done by applying geophysical methods during different stages of exploration (Bhimasankaram, 1975). There has been a number of studies to explore the groundwater resources using geoelectrical investigations (Yadav and Lal, (1989); Kshirasagar and Rao, (1989); Kishore and Bhimasankaram,(1982); and Yadav and singh (1985). The Upper Godduvanka river basin rises from the gamkonda hill ranges and flows in north east direction joining in Ponnai river. The Upper Godduvanka river basin has an areal extent of 478 Km^2. The basin lies in between north latitude 13° 8' - 13° 23' and east longitude 78° 43' 15" - 78° 59' 50". The study area includes the sub-basins of Bahu, Penrevulaeru and Bangarupalemeru.

MATERIALS AND METHODS

71 vertical electrical soundins (Fig.1) were carriedout using wenner configuration . A D.C. earth resistivity meter was used. The maximum lengths of current electrode separation (a) were kept between 90 and 100 m from place to place depending upon the accessibility of the area. The data was interpreted by partial curve matching with the help of master curves. The resistivity data is analysed qualitatively and quantitatively to delineate the groundwater resources of the study area.

V. P. Singh and B. Kumar (eds.), Subsurface-Water Hydrology, 211–221.

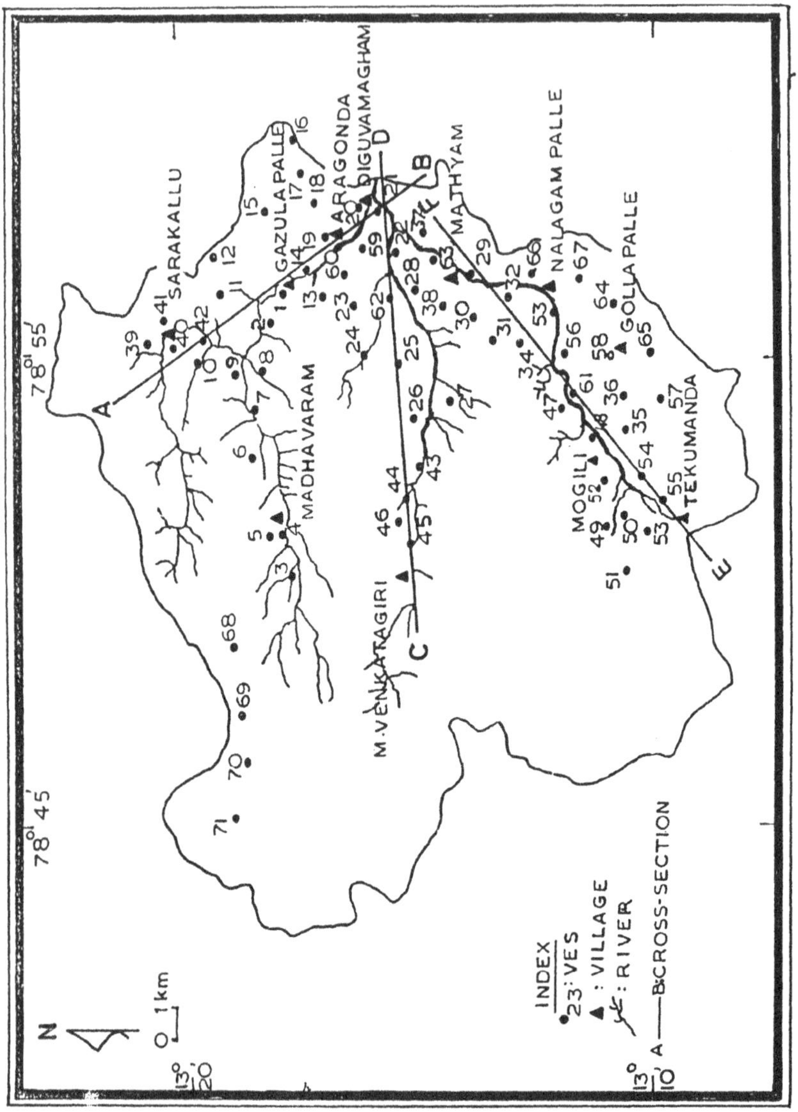

FIG.1 Location map of vertical electrical resistivity soundings

RESULTS AND DISCUSSION

QUALITATIVE INTERPRETATION

The variation in apparent resistivity in the area often can be related qualitatively to the geological features (Schwartz and McClymont, 1977). In qualitative interpretation, the contour map of apparent resistivity for the separation a = 30 m is prepared (Fig.2) with a view to delineate high and low resistivity zones and nature of the basement rock. Based on the field conditions, the resistivity value of 100 ohm-m for separation a = 30 m is approximately selected to delineate the good and poor water bearing areas. The areas having lower resistivity, outline the channels along which groundwater appears to flow the shear zones and low lying areas. Most of the wells yielding comparitively good quantity of water are located in low resistivity zones. The high resistivity contours are located over topographic highs. Most of the wells located in high resistivity zones are giving relatively poor yields.

QUANTITATIVE INTERPRETATION

The field curves have been interpreted using the curve matching method with the help of available two and three layer master curves and auxillary graphs (Orellana and Mooney, 1966). The following are the characteristic resistivities obtained for various formations based on sounding interpretation.

	Resistivity	Thickness
Soil	5 - 140 ohm-m	1 - 5 m
Sand	8 - 30 ohm-m	5 - 24 m
Weathered granite	10 - 100 ohm-m	4 - 24.5 m
Fractured granite	125 - 450 ohm-m	7 - 40.8 m
Hard rock	>500 ohm-m	

The shape in most of the curves are A,H,K and KH types. In general the curves are of four layers. The A type curves indicate that the low resistivity soil cover overlying the weathered granite and hard rock. In H type curves, the soil behaves as a highly resistive layer as compared to the weathered layer. The H type curves indicate the presence of thick layer of alluvium consisting of find sand to sandy gravel. K type curve indicate weathered rock below the soil cover followed by fresh granite. In all the cases, the terminal branch of the curve rises linearly at an angle of 45° indicating the infinitely resistive bedrock.

As mentioned earlier, the top layer is soil followed by sand/weathered rock underlied by either fractured or compact granite. The soil cover thickness varies from 1 to 5 m with resistivity variation of 5 to 140 ohm-m. The thickness of alluvium which consists of gravel, sand and silt varies from 5m (S-1) to 24 m (S-8), whereas its resistivity range varies from 8 to 30 ohm-m. The thickness of weathered rock varies from 4 to 24.5 m with resistivity ranging from 10 to 100 ohm-m followed by hard basement with very high resistivity value (500 ohm-m). In a few curves, whose resistivities exceed 100 ohm-m (having a range of 125 to 450 ohm-m) represent a fractured granite.

The alluvium with considerable thickness (24 m) having resistivity range 10 to 30 ohm-m acts as a good aquifer due to its permeable nature. Alluvium having 10 ohm-m resistivity range can be considered as semi-permeable zone, as it is likely to consist of clay particles. The weathered granites are also proved to be good aquifers under water table and artesion conditions. The weathered granite with 14 m thickness with resistivity range 10 to 60 ohm-m are considered to be good groundwater potential zones.

VERTICAL GEOELECTRICAL CROSS SECTIONS

Vertical geoelectrical cross sections are prepared based on the results of vertical electrical soundings data to study the variations in thickness of different layers, i.e., soil, sand, weathered and fractured granite in sub-surface along three profiles AB,

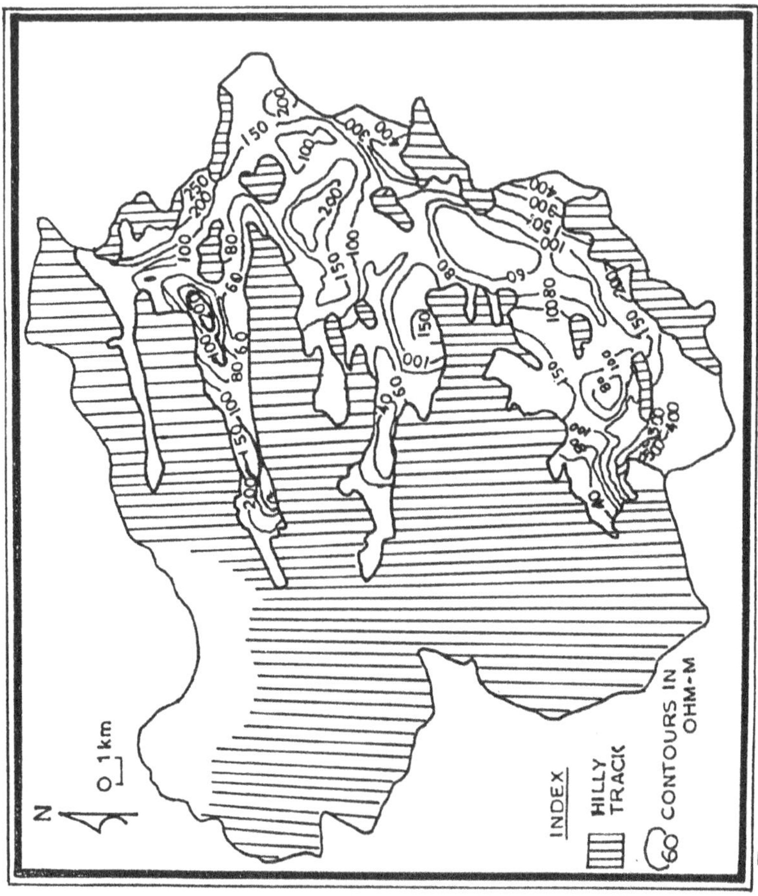

FIG.2 Apparent resistivity contour map for a=30m separation

CD, EF (Fig.3). The ground levels of sounding points are not taken into consideration while preparing the cross sections. For the sake of clarity, resistivity values are given in their respective positions.

Figure 3a shows the cross section AB (NW-SE) between Sarakallu and Diguvamagam based on the results of six soundings (1,14,19,21,40 and 42) data. The depth of inference between soil and sand determined from the geoelectrical soundings, is in good agreement with that obtained from the well inventory data. Soil cover varies from place to place and maximum upto a depth of 9 m at S-21 and minimum upto a depth of 1.8 m at S-1. Sand zone occur all along the profile. Maximum depth of sand zone upto 15.5 m at S-40 and minimum depth upto 9 m at S-21. Its resistivity varies from 11 to 51 ohm-m. Sand zone thickness decreases from NW to SE side of the profile. Weathered granite is present like a valley between S-42 and S-21 in the AB profile with the exception of S-40. Weathered granite upto a maximum depth of 32 m is present at S-1 and minimum 13.5 m at S-42. Fractured zone is found only at S-14 upto a depth of 40.8 m. In S-42 hard rock is likely to be found at 15 m. Dugwells are recommended on the edges of the profile, due to the presence of shallow basement. Good groundwater potential zones are likely to occur in between S-1,S-14 and S-19, due to the presence of sand, weathered and fractured granite.

A second cross section CD (W-E) between M. Venkatagiri and Diguvamagam based on the results of seven soundings (21,22,25,26,44,45 and 62) data (Fig.3b). In this profile, soil cover thickness varies from place to place and its maximum depth upto 6 m at S-21 and minimum depth upto 1.2 m at S-25. Sand zone occur all along the profile with exception of S-22, where it is absent. Sand zone thickness decreases from west to east side of the profile with maximum depth upto 12.5 m at S-45 and minimum depth upto 6 m at S-26 and its resistivity ranges from 8 to 35 ohm-m. Beneath the sand zone, weathered granite occur all along the profile except at S-25. Maximum depth of weathered granite occur on the edges of the profile (S-22,S-62,S-44 & S-45) and it is completely absent at S-25. It occurs like a dome shape in the profile. Fractured zone is likely to found only in S-26 upto a depth of 29.2 m and is absent in the rest of the profile. Hard rock is found to be shallow at S-25,S-21 and S-44. Good groundwater potential zones likely to occur in S-45,S-62 and S- 22, due to the presence of weathered and fractured granite. Along this profile, wells can be recommended at S-45, S-62 and S-22. The thickness of weathered rock varies from 3.2 m to 25 m, with a resistivity range from 15 to 100 ohm-m, whereas the fractured rock occurs for a maximum thickness of 20 m at S-26 with a resistivity of 11 ohm-m.

A third cross section EF (SW-NE) between Tekumanda and Mathyam based on the results of six soundings (29,32,34,54,55 and 61) data (Fig.3c). This EF (SW-NE) section lies approximately perpendicular to AB (NW-SE) cross section. Soil cover occurs all along the section and its thickness varies from place to place. In this section, the occurrence of the sand zone is like a valley shape. The maximum thickness of sand zone including soil cover is 12 m (at S-61) in the central part and minimum depth 6 m (at S-55) in edges of the section. Weathered granite is completely absent in S-61 and its thickness increases towards edges of the section. The weathered granite occurs upto a maximum depth of 15.4 m at S-34 and the resistivity of the weathered granite ranges from 35 to 100 ohm-m. Fractured zone is completely absent through out the section. Hard rock is likely to be found at shallow depth along the section. Degwells alone are recommended along the section, due to the presence of hard rock at shallow depth. In this section, it is found that there is no feasibility to tap groundwater from the deeper aquifers.

CORRELATION OF SOUNDING RESULTS WITH GEOLOGICAL SECTIONS

The results of a few resistivity soundings taken close to the wells have been correlated to the available geological sections (Fig.4). It is clear from the figure 4 that the thickness of soil as derived from the interpretation of sounding curves is in good agreement with the actual geological section.

In most cases in alluvium area, the boundary between the second layer (sand) and the third layer (weathered granite) of the geological sections is not so well

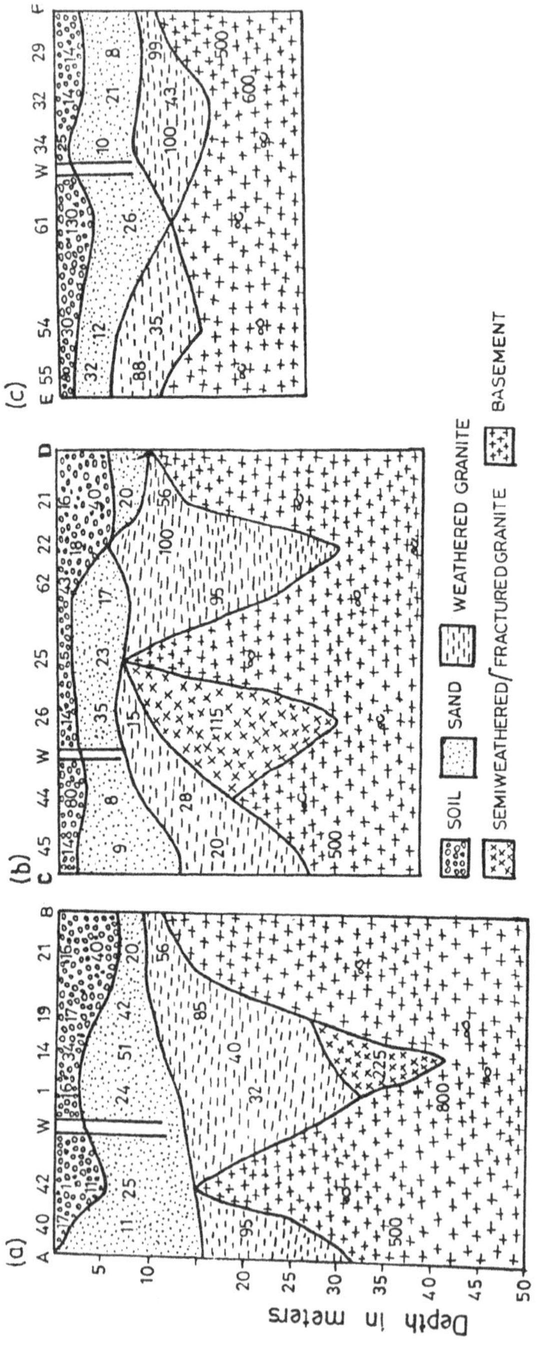

FIG.3. Vertical geoelectrical cross sections along profiles (a) AB (b) CD (c) EF

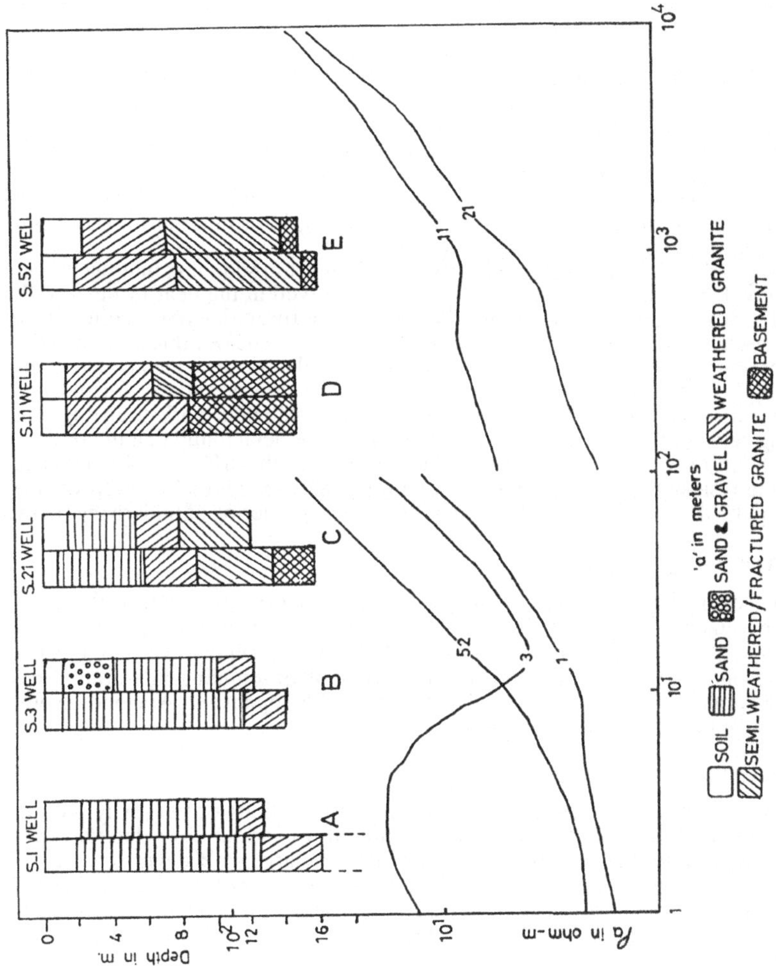

FIG.4. Correlation of resistivity soundings with Geologic sections of Existing Wells

brought out by the resistivity soundings. Eventhough the geological sections indicate the presence of four layers i.e., sand, weathered rock, fractured rock and hard rock, the interpretation of the soundings shows only three layers instead of four (Fig. 4b). This indicates that the alluvium thickness as interpreted from soundings includes partly, the highly weathered granite layer as well. So the four layers are electrically equivalent to three layers, which may not be having sufficient resistivity contrast so as to bringout the difference between them. At places, in figure 4a the interpreted thickness of sand is more than the actual thickness observed in the nearby well as same part of highly weathered granite offered similar resistivity of sand. But in case of S-21 (Fig.4c) the lithology derived from sounding interpretations are in good agreement with the actual lithology of the well.

In granite area, the geological sections derived from sounding interpretations are in good agreement with the actual sections observed in the near by open wells S-52 (Fig.4e). But in few cases as in S-11 (Fig.4d), interpretation of soundings show only three layers instead of four as the interpreted weathered layer thickness includes partly fractured layer. It shows that the four layers are electrically equivalent to three. The depth to the basement derived from sounding interpretations are in good agreement with the actual depth data, collected during the field work from dugwells and dug cum borewells. From these observations, it has been found that the reliable information regarding the electrical properties of the sub-surface can be obtained using the resistivity surveys. The electrical properties can, inturn, be interpreted in terms of the degree of water saturation or degree of weathering of underlying rocks.

DEPTH TO BASEMENT

Based on the results of soundings, a map has been prepared showing the depth to the basement (Fig.5). It is clear that the depth to basement is not uniform in the basin area, the depth being maximum at S-14 (40.8 m) and minimum at S-16 (4.4 m). As mentioned previously on the basis of interpretation of sounding data alone, areas around Sarakallu, Mallakunta, Gazulapalle, West of Kuppam and Aragonda seems to have a good groundwater potential zones.

GROUNDWATER POTENTIAL ZONES

As it is evident from the quantitative interpretations of resistivity data of soundings, it may not be correct to expect continuity of various layers laterally over a large distance in the basin area. Believing that very large scale deviations are not present within a few kilometers, groundwater potential zones have been delineated and presented in figure 6. This map is based on the apparent resistivity contour map, thickness of the aquifer and depth to the basement obtained from quantitative interpretations.

Areas having 12 m thickness of sand with resistivity range 10 to 30 ohm-m, and 14 m thickness of weathered granite with 10 to 60 ohm-m, are considered as good groundwater potential zone, whereas the area with 14 m thick weathered granite having resistivity range 60 to 100 ohm-m is considered as fair groundwater potential zone. The area having a thickness (weathered granite) of less than 14 m and having resistivity range above 100 ohm-m is considered as uncertain groundwater potential zone.

It is concluded that the groundwater potential zones derived from qualitative interpretations (apparent resistivity contour map) are considerably in good agreement with the groundwater potential zones derived from quantitative interpretations.

CONCLUSIONS

The overall study reveals that the groundwater table is found approximately at 7 m depth. The exploitable groundwater is found in the sands and weathered granite layers at the varying depths between 15 to 25 m especially in north, west, southwest and southeastern parts and rest of the area at shallow depths. Fracturing is found to be more with depth at central and southern parts which contains sufficient resources

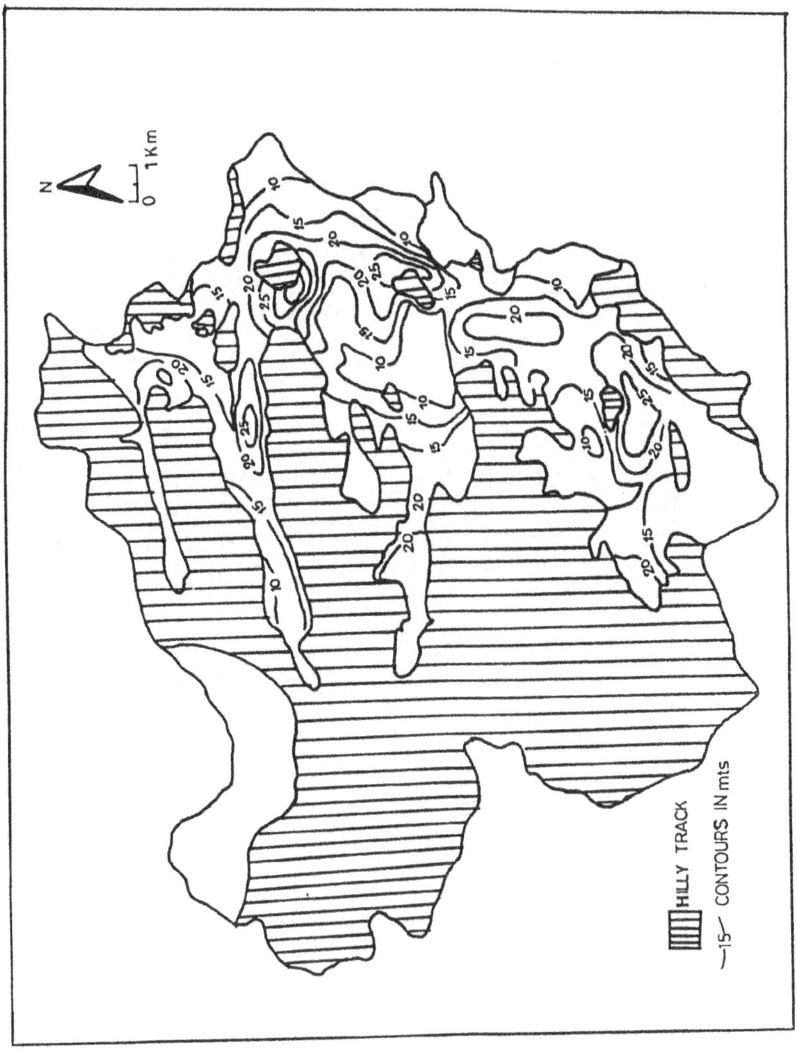

FIG.5. MAP SHOWING INFERRED VARIATIONS IN DEPTH TO BASEMENT

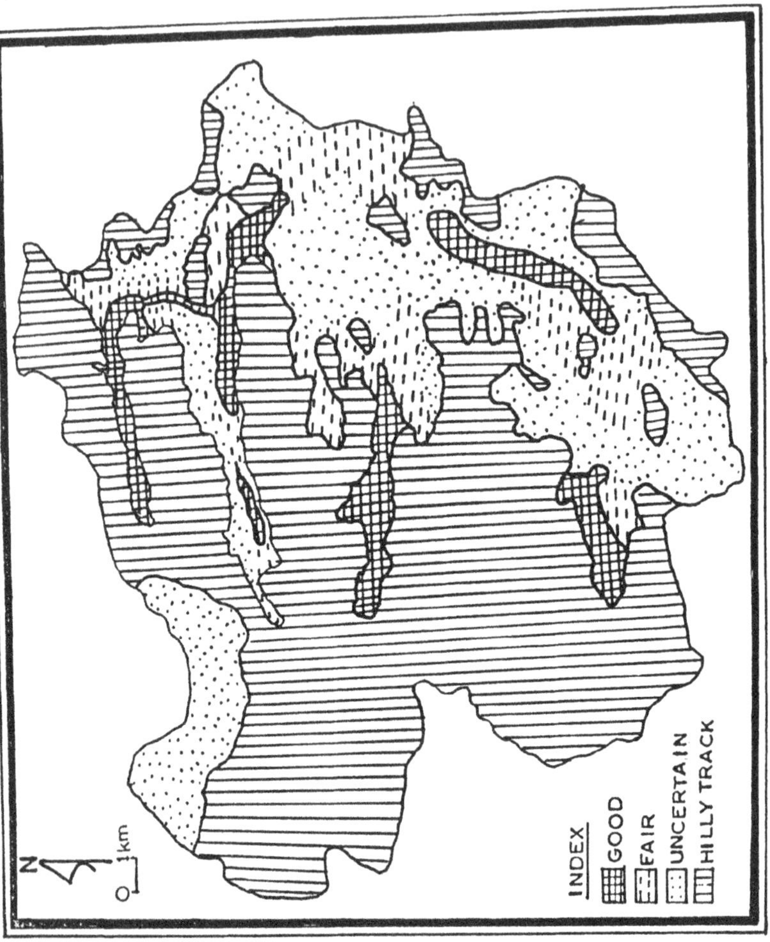

FIG. 6 Map showing different groundwater potential zones

of groundwater and it could be exploited at specific points from depths ranging from 25 to 40 m. Based on the depth to basement map, it is found that the basement is not uniform in the study area and it varies from 4.4 to 40.8 m. Another important finding is that the good groundwater potential zones are present along the foot of the hills and valley portions (north, west and southwest), and in southeastern parts along the weathered and deep fractured granite rock

REFERENCES

Bhimasankaram, V.L.S. 1975. 'Application of electromagnetic and Scismic methods for groundwater exploration in hard rock areas'. Paper read at *Workshop on groundwater exploration in hard rock terrains*. NGRI, Hyderabad.

Kishore, B.H.B. and Bhimasankaram, V.L.S. 1982. 'Analysis and evaluation of groundwater potentialities in the granitic aquifer system of Shadnagar basin', *Geophysical Research Bulletin*. **20**(4), p 173-181.

Kshirasagar, T.V.S.R, and Nagamalleswara Rao, B. 1989. 'Electrical resistivity survey for groundwater in Varaha river basin, A.P., India'. *International workshop on appropriate methodologies for development and management of groundwater resources in developing countries*, NGRI, Hyderabad. **1**, p 329-332.

Orellana, E. and Mooney, H.M. 1966. 'Master tables and curves for vertical electrical soundings over layered structures'. *Intercieneia*, Madrial, Spain.

Schwartz, F.W. and McClymont, G.L. 1977. 'Applications of surface resistivity methods'. *Groundwater*. **18**(3), p 197-202.

Todd, D.K. 1959. *Groundwater hydrology*. John wiley and sons. New York.

Yadav, G.S. and Singh, C.L. 1985. 'New sets of direct and inverse schlumberger filters for resistivity sounding interpretation'. *Bolletino di Geofisica Teorica ed Applicata*. **27**(108), p 251-262.

Yadav, G.S. and Lal, T. 1989. 'Investigation of groudwater resources at selected levels in drought prone area of Pahari block,District Mirzapur, U.P., India'. *International workshop on appropriate methodologies for development and management of groundwater resources in developming countries*, NGRI, Hyderabad, 1, p 209-219.

EXTRACTION OF DRAINED WATERS FROM INSTABLE AREAS FOR MINOR AGRICULTURAL USES

SPILOTRO G.^, VACCA G.°
° Department of Environmental Engineering and Physics
^ Department of Structural and Geotechnical Engineering and of Engineering Geology
BASILICATA UNIVERSITY - POTENZA (ITALY)

ABSTRACT. Water resources of low potential, generally associated with low permeability soils, are of considerable interest especially in many Italian bad lands. Their extraction and use are very important for the following reasons:

- these resources are usually associated with instable areas and in the majority of the cases they are responsible for the instability. So drainage may be the most important system for the reclamation of these areas.
- although modest, these resources may be of interest for small local agricultural use.
- they may be utilized at a reasonable cost even in extremely peripheral areas, not easily reached by the energy supply.

In the paper the characteristics of low permeability aquifer systems, will be examined and described. The advantages of the improvement of the stability of the slopes in these areas after drainage will be highlighted.

Finally the usefulness of the exploitation of aeolic energy for the withdrawing of drained waters using new technology, will be demonstrated. The climatic zone of these areas assures the availability of wind resources for most of the year and particularly during bad weather seasons.

1. INTRODUCTION

Major water resources have been the focus of attention in countries with the most ancient civilizations for a long time; their use in civil and agricultural projects is in some cases centuries old, while it is only in the last few decades that they have been included in more complex water schemes.

Water resources of minor potential, associated with soils of modest permeability and or with aquifers of small dimensions, may be of interest from different aspects.

- one is dealing with a resource which may be of local interest.
- in particular morphological contexts it is a resource generally associated with instable areas.

Such interest may be sensitively revalued, where the aquisition of the resource at low cost in the area where it is available proves possible. The traditional techniques of drainage by

V. P. Singh and B. Kumar (eds.), Subsurface-Water Hydrology, 223–230.

gravity or by means of mechanical withdrawal using the electricity network are often vulnerable and expensive.

The solution may be found in many cases by resorting to traditional techniques of utilization of alternative energy, applying new technological evolutions.

2. LOW PERMEABILITY AQUIFER SYSTEMS IN ANTHROPIZED ENVIRONMENTS

Low permeability aquifer systems in some widespread areas of southern Italy, for example the marine terraces of the bradanic area and of other adjacent areas which are geologically similar, were already well known from the time of the greek colonization and they influenced the localization of the first organized settlements. The availability of water, either from springs or wells, was a determining factor in this localization and of the successive organization of such settlements into inhabitated centres.

The drainage operated by anthropic activity was quite important in the control of underground flows and contributed in an often not unimportant way to the control of instability in these inhabited settlements, which were linked for reasons of defence, at the summit of the slopes.

The arrival in these zones in this century of aqueducts, and, successively of sewage networks, implicitly caused a worsening in the conditions of stability, provoking the feeding of the groundwater. This was caused by the failure to withdraw water from the aquifer and the fact that waste deriving either from the water network or the sewage network was immitted directly into in the groundwater.

The cause and effect relationship cannot be certified in many cases, because of the lack of registration of relevant information regarding the groundwaters before or after the advent of the aqueducts; however it has been historically certified that the instability of peripheral areas of many inhabited centres, before the huge urban extensions of the middle of this century, underwent notable increases after the advent of water systems.

3. LOW PERMEABILITY AQUIFERS

Soil containing water, independently of the specific quantity contained, may be considered an aquifer if it is possible to separate the water from the soil containing it by means of a work of force. If the force is gravity this is not sufficient in fine grained soils of low permeability to overcome hydrodynamic resistance and the forces of molecular attraction.

Therefor systems characterized by soils with grains of inferior dimensions to those of fine sands, with silt and clay contamination, cannot qualify in general terms as aquifers, even though theoretically one is dealing with saturated porous media.

Actually in our region almost all the natural soils, including sedimentary with fine grains, belong to structured masses, following singenetic or post depositional processes.

In such masses significant small or large hydrogeological structures can occur, which allow the presence and circulation of free waters. The paths are not controlled only by variations in permeability of stratigraphic and tectonic lithological sequences.

THe most important structures of hydrogeological interest which can be discovered in low permeability soils are:

- granulometric or lithologic variations deriving from normal sedimentary processes
- granulometric variations along horizontal plains, deriving from the superimposition of landslide detrital deposition.
- granulometric variations deriving from processes of surface alterations
- the removal of more soluble minerological components (eg. carbonates)
- precipitation and deposition of chemical compounds
- normal sedimentary discontinuity: stratified surfaces
- fracturing of undetermined origin
- localized fracturing of either tectonic or mass movement origin
- scaliness (of tectonic origin)
- phenomena such as chemical dissolution or localised erosion: pseudo-phenomena or para-phenomena of karst (GUERRICCHIO et Al.,1979)

The practical consequence is however linked to the presence of free water in such masses, which permits the detection of a groundwater surface, the reconstruction of a three-dimensional flow net and, finally, possible water catchment.

4. SOIL INSTABILITY IN LOW PERMEABILITY AQUIFER SYSTEMS

The conditions of stability in low permeability morphologically not laterally confined aquifer systems (valley slopes, coastal cliffs, etc.) are generally bad. This is due to at least three factors:

- because of the low permeability, the piezometric heads remain high even near the surface of the slope
- because also the soils are partially deprived of lateral confinement, in the presence of water all the processes of the swelling of the underlying consolidated clays and the alteration of other types of soil are favoured
- the elevated gradients of filtration close to the terminal lines of the flow may provoke, also for the reasons specified above, direct and regressive erosive phenomena.

The instability phenomena which derive from this, in the first instance are generally of a circular generatrix (landslides) and may have very extended dimensions. Typical evidence is derived from the formation of small ponds behind counterslope landslide terraces. The hydrogeological disorder which follows the phenomena of the first failure, leads to further evolution of instability phenomena, which affect soils which are by now structurally destroyed. The final evolution is in the form of flows of earth or mud.

In landslides with a prevalent longitudinal extension (FIDELIBUS et Al, 1990) mechanisms of the circulation of water in the body of the landslides have been recognized, which are reputed to be responsible for the reactivation or the phases of quiescience of same.

In each case it is evident that the necessary condition (even if it is not always sufficient) for the achievement of stabilization in such areas is the drainage. In fig.1 we show in the case of an infinite slope in cohesionless soil the effect on safety factor of the groundwater lowering.

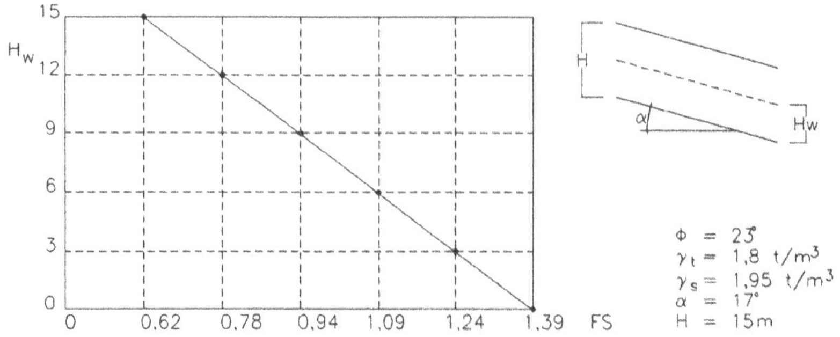

Fig.1 Safety factor vs Hw for an infinite slope in cohesionless soil

5. EFFECTIVE INFILTRATION IN LOW PERMEABILITY AQUIFER SYSTEMS

Valuations relative to the bradanic area (COLELLA et Al., 1993) , stratigraphically characterized by the superimposition of conglomerates and sand, often cemented, on clays of notable thickness, highlight, compared with a medium rainfall in the range of about 700 mm/y, effective infiltration values which reach about 60 mm/y . Such effective infiltration feeds both the waters at the contact between more permeable soils and clays, and water circulation within the structures of the same sandy and clayey mass. Finally, both aquifer systems, of which evidently the first constitutes up to the point of exhaustion of same, superior border conditions for the second, feed the movement of water by capillarity towards the topographic surface of the system, where evaporation and/or evapo-transpiration takes place.

Such effective infiltration leads to a potential annual water storage of 60.000 m^3/Km^2.

6. DRAINAGE OF LOW PERMEABILITY AQUIFER SYSTEMS

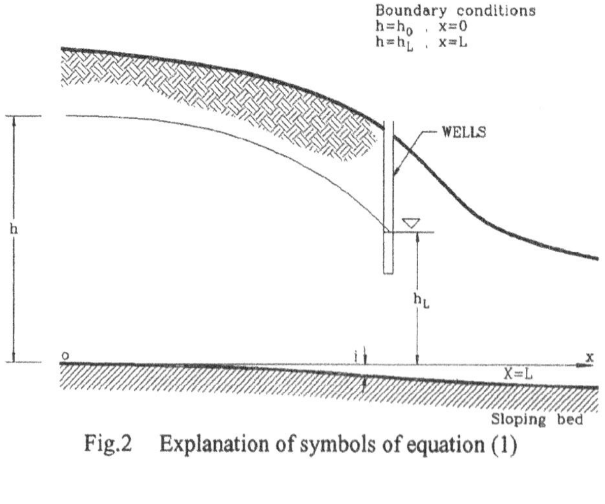

Fig.2 Explanation of symbols of equation (1)

For practical purposes, the aquifer to be drained can be modeled as an isotropic homogeneous medium lying over a less permeable bed with a top surface having an i slope towards the old plio-pleistocenic coastal line and intercepted by the river valley.

A continuous draining diaphragm (or a wells line) (fig.2) lowers the groundwater level from its initial undisturbed level h_0 to a new level h_L under a continuous discharge Q. We have steady state conditions only if we can find at distance L from the draining line a source that holds undisturbed the h_0 level (BEAR, 1979). On the other hand, this is possible only if the discharge is not greater than the rate of flow of the water in the aquifer. Otherwise we must concern ourselves with aquifer storativity and then with non steady conditions.

The relationship between the drawdown and the discharge is done by the non linear equation:

$$Dh = h_o - h_L = Q*\ln((h_0+Q/(k*i)) \;/ (h_L - i*L + Q/(k*i)))\;/(k*i) \qquad (1)$$

A typical result is shown in fig.3 . We can observe that the same discharge can be reached with several drawdowns under different border conditions. This arises because the border condition (or the radius of influence in the case of a well) can be evaluated mainly by empirical observations.

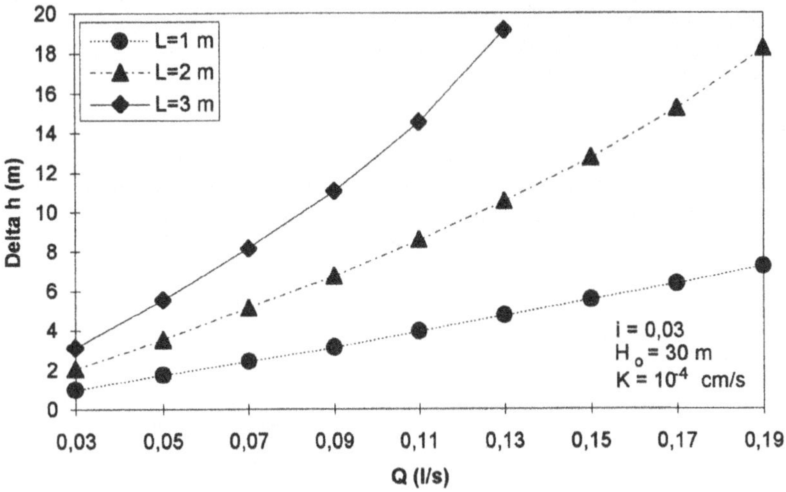

Fig.3 Discharge from a vertical drain for unit lenght

To improve the efficiency of the drainage in this type of soil it is necessary to widen the difference $D_h = h_0 - h_L$. This however is a notable obstacle to natural gravity drainage, from wells of a diameter so large as to allow the execution from within of the perforation of waste to the outside. With the increase of Dh the length of such conduits increases, which are not esasily executed and are instead very vulnerable. These infact cross in their terminal tract instable soils, which with slight movements can break the conduits and render useless very expensive drainage works.

The problem can be more easily and economically resolved with mechanical drainage from wells of normal diameter, given that the lifting energy is available at low cost in the area. Electrical networks are also very expensive and, like the waste conduits, are vulnerable in that they have to cross instable areas.

For the subsequent computations we assume a discarge of 0.1 l/s with interspacing of the wells of about 3 m and with drawdown of 15 m starting from a groundwater level of 15 m .

7. THE UTILIZATION OF WIND ENERGY

Wind power for centuries provided most of the energy human requirements by means of mills; over 8000 years ago it was already used to drive sail boats. With the advent of more convenient forms of energy, its use has diminished in the United States as well as in Europe in the years to an insignificant fraction . However, with impending shortage of oil and natural gas and mostly with steeply rising energy prices, in the 80's there was a renewed interest in all renewable energy sources, wind power included.

Most of the aerogenerators are horizontal axis machines today, equipped with electric generator or directly connected to any utility. The wind power increases as the cube of the wind speed and when the generator rating is reached, the power is regulated to this value for all higher speeds by blade pitch control for example.

Simple and efficient applications in agricultural field are very common in the States as well as in Europe, especially in its Northern part.

In this context, the mechanical drainage from wells appear efficient and economically appropriate if aerogenerators are used to provide for energy consumptions.

Apennine and Sub-Apennine regions in Southern Italy have normally a well distribution of wind resource during all seasons, particularly in autumn and winter, which are moreover the most rainy seasons. Figures 4 and 5 represent the wind velocity distribution for one site in the South-eastern Apennine (Daunia hills) .

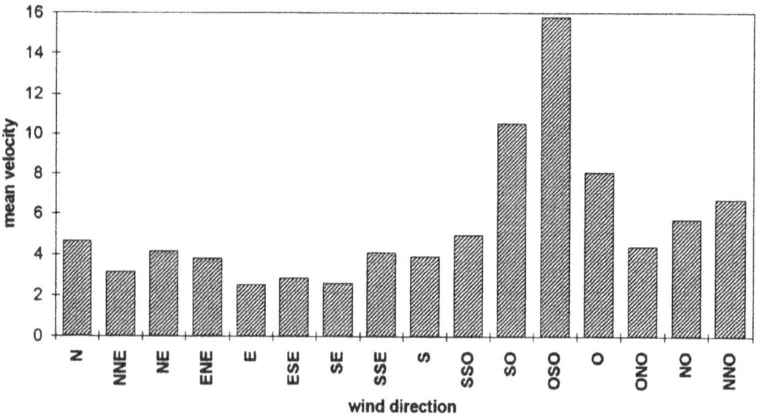

Fig. 4 Wind mean velocity distribution Nov. 1992 - Jan. 1993

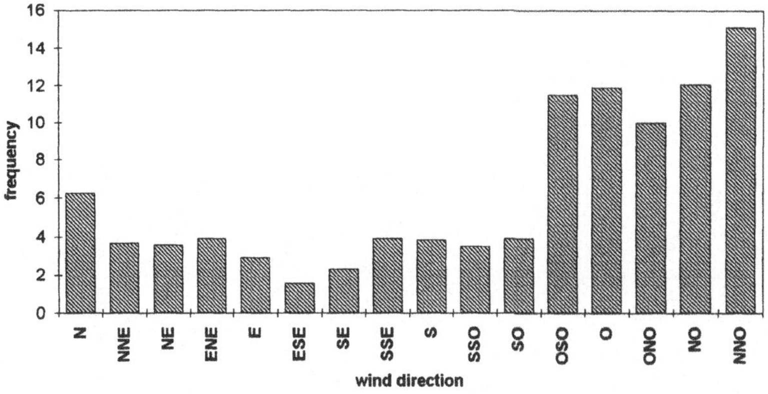

Fig.5 Wind velocity frequency Nov 1992 - Jan 1993

On the basis of such distribution, a single blade (NACA 44xx series) small wind turbine is usable (5.2 kW rated power for a windspeed of 10 m/s) , operating in the range 3.9 - 20 m/s and able to drive electrically a canned pump (2.2 kW rated power at 50 Hz) to discharge up to 20000-25000 m^3/y of water with a mean wind velocity of 6 m/s and a mean head of discharge of 50-60 m . A typical performance curve of the aerogenerator is given in fig.6 .

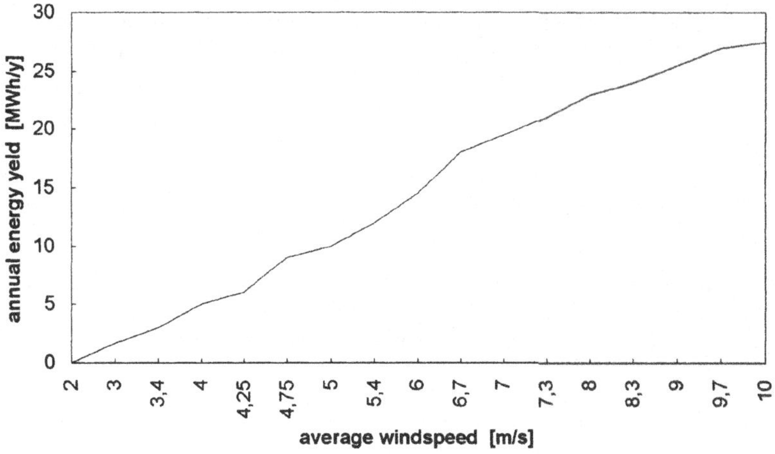

Fig. 6 Wind turbine performances

In the range of low permeability soils we have already estimated that a discharge of about 0.1 l/s (9 m^3/d) can be obtained with a drawdown in the range of 10-15 m . In this case a single wind turbine is able to drive up to six pumps, with single commutation and direct connection (without energy storage) .

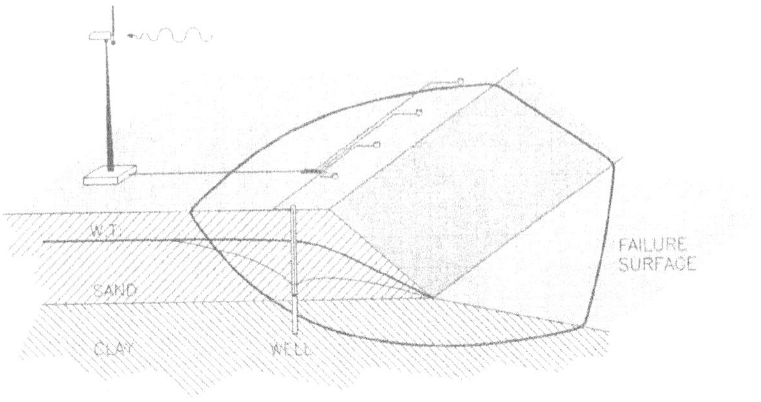

Fig.7 Layout of drainage system of a landslide area by means of wells with pumps energized by aeolic generators.

The diameter of the well determines a local water storage capacity, that can be used to control the time lap of the electrical connection of the single pump in the well.

The overall daily recoverable discarge is then in the order of about 50 m^3/d ,also available during the dry season. The cost of the plant is intended to be payed for land reclamation purposes, in such a way that it will be available at zero cost for minor agricultural uses all over the year. Small storage basins, for which also new tecnologies are now available, can be utilized to optimize the use of these waters.

REFERENCES

Bear J. (1979) Hydraulics of groundwater
Mc Graw Hill, N.Y.

Colella M., Grimaldi S., Sdao F., Spilotro G. (1993) Idrogeologia del settore lucano della Fossa Bradanica: analisi in un'area campione.
Conv. Naz. Giovani Ricercatori in Geologia Applicata, Potenza ott. 1993

Guerricchio A., Melidoro G. (1981)
Geologia Applicata e Idrogeologia, vol.xx ,1981

Spilotro G. (1992) Recupero di acque di drenaggio di aree instabili per usi agricoli minori.
1° Int.Environmental Workshop on Integrated Water and Wastewater Management and Integrated Solid Waste Management. Politecnico di Bari & State University of New York, Bari nov.1992

Fidelibus C., Lenti V., Sdao F., Spilotro G. (1990) Condizioni strutturali ed al contorno nell'attivazione di movimenti di massa a prevalente sviluppo longitudinale.
Proc. Meeting on: Runoff Processes and Morphological Evolution of Slopes. Univ. of Basilicata, may 1990

GROUNDWATER HYDROLOGY OF SIWANA, WESTERN RAJASTHAN -- A MULTIDISCIPLINARY APPROACH

H. Chandrasekharan, M.L. Sharma* and Y.P. Singh**
Nuclear Research Laboratory,
Indian Agricultural Research Institute, New Delhi-11012

ABSTRACT Studies in groundwater hydrology always need a multidisciplinary approach for their understanding and evaluation. Although there are a number of literature dealing with different aspects of groundwater hydrology in detail by individual technique/approach; it is often observed that an integrated approach is lacking in most of them. In this paper, we discuss the groundwater hydrology of Siwana, an intermontane basin, western Rajasthan, wherein use of hydrogeological, geophysical, geo-chemical and isotopic investigations have been incorporated and discussed.

Studies reveal that the aquifer in the southern part of Siwana are associated with clay and silt. Clay lenses are also found at shallow depth causing negligible vertical recharge to the groundwater as shown by the injected tritium profiles. Stable isotope contents of deep dug wells in the area indicate that the groundwater are of meteoric origin but has not undergone evaporation. Geo-chemical investigations coupled with the results of other geohydrological investigations indicate saline water zones north of Siwana and potable water at other locations. A general picture of groundwater conditions has been presented on the bases of results of different geohydrological investigations.

* Central Arid Zone Research Institute, Jodhpur-342 003
** Defence Laboratory, Jodhpur-342 001.

V. P. Singh and B. Kumar (eds.), Subsurface-Water Hydrology, 231–243.

Introduction

The Thar desert has been afflicted by recurring droughts throughout ages; the root cause being problems associated with water resources. Many organisations including the internationals like UNDP, UNESCO, WHO, ETC have carried out multidisciplinary approach for ground water studies. The phase II of the IGCC project, when completed, is expected to cover about 10.6 per cent of the arid region under canal irrigation. The ephemeral river Luni system with 79 minor and major dams covers about 16 per cent area. But the remaining 73 percent are principally dependent on ground water resources for their subsistence and agricultural development. Out of this, about 62 per cent area is covered under dunes and blown sand of varying magnitude obsecuring the study of concealed geological formations. In addition to the above, the gradual growth of population of man and animal and lack of suitable surface water sources increase more stress on the groundwater usage causing indiscriminate over-exploitation. This obviously results in the lowering of static water level, reduction in well yields and increase in total soluble salt contents. The annual demand for groundwater has increased from 2000 mcm to 4150 mcm in the last 20 years (1960--1980) (Mehra, 1984) with the projected value of over 7000 mcm in the early nineties. It has been reported that groundwater in about 65 per cent of the area in western Rajasthan has salinity of over 3200 ppm (Chatterjee and Vengni, 1984).

Assessment of groundwater under the above background needs an integrated approach involving differnt disciplines of geo-sciences. One such investigation conducted around Siwana, an intermontane basin in Barmer district, western Rajasthan, India, are presented and the results are discussed in the light of other available information.

Materials and Methods

THE STUDY AREA

The area of investigation, Siwana (Fig. 1) is well within the eastern boundary of the Great Indian Desert. Here, the outcrops dip towards the centre from all around and the slope is towards the west on the southern side and towards north on the northern side. An ephermeral river called, 'ver nadi' originates from the hills near Piplun (south of Siwana) and disappears near Indirana, about 20 km west of Siwana, due to alluvial suffocation.

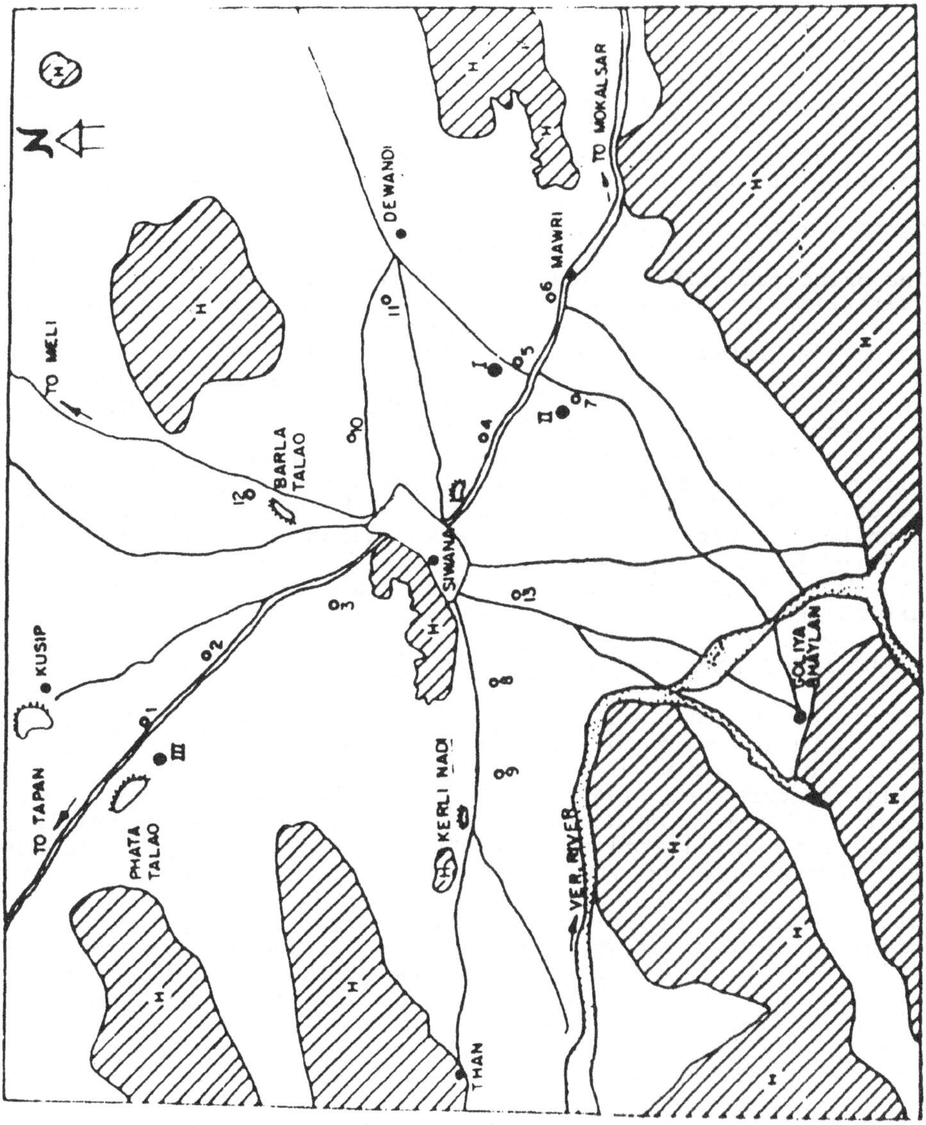

Legend: O -- VES Point as well as Ground Water Sample
● -- Tritium/Cobalt-60 Injection Site
II (Of the three injection sites, only two are reported in this paper)

Fig. 1 Map of Siwana, Western Rajasthan, showing Locations of Different Investigations

The area receives the annual rainfall of about 30 cm average for 84 years from 1901) and ranges from 2.67 cm in 1918 to 101.8 cm in 1926. Nearly 90 per cent of the rainfall occurs during the southwest monsoon; that is, July--September). The mean maximum and minimum temperatures range from 25°C to 43°C and from 5°C to 28°C respectively. During summer, dust storm (Andhi) and dry hot dust wind (Loo) occur frequently. The annual evapotranspiration is of the order of about 200 cm.

GEOLOGY AND HYDROGEOLOGY

The principal lithological formations of Siwana are alluvium of quarternary, older alluvium, gruss of Pleistocene (Chatterjee, 1963) and Malani volcanics, Siwana and Jalore granites of Paleozoic (Lower Vindhyans). Jalore granite outcrops occur in a limited area. They are red to pink in colours, porphyritic, hard, massive and compact in nature and consists mainly of biotite (Adhyalkar, 1964). Of all lithological formations, calcareous blownsand and younger alluvium are the aquifer with the greatest water potential. Primary rocks of this region like volcanics and granites are poor hosts for water accumulation except in joints, bedding planes, cracks, fissures and the weathered mantle. The depth togroundwater is in the range of 22--30 m. below ground level.

FIELD INVESTIGATIONS

Thirteen vertical electrical soundings (VES) of Schlumberger configuration were conducted in a close grid pattern (Fig 1). Field data were interpreted using standard master curves (Rijikswaterstaat, 1969) and the interpreted data were verified in a computer. Dar Zarrouk parameters (Mallet, 1947) were calculated as S and T for 30 m. column thickness and presented along with the resistivity in Fig. 2 (a & b) and Table 1. Radioactive tracers were used to estimate the ground water recharge, if any, at two places (with tritium tracer at both and cobalt-60 at one place only) beside the sounding points. After analysing the samples collected two/four times, respective histograms of tritium and cobalt-60 profiles are shown in Figs. 3 a,b, and c. Ground water samples from deep dug wells adjacent to the VES points were also collected for stable isotopic analysis. These data are represented in Fig.4. The soil moisture contents, in-situ bulk density, electrical conductivity, etc., were also estimated by conventional methods.

Results and Discussion

Resistivity data (Table 1) indicate that the surface layer

Table 1 Resistivity Data Around Siwana, Rajasthan

VES No.		LayerI	LayerII	LayerIII	LayerIV	LayerV	T(30 m.	S(30 m.)
1	h	1.0	2.0	8.0	50.0	?	770	1.628
	p	100.0	114.0	25.0	15.0	76.0		
2	h	1.0	3.0	15.0	60.0	?	345	4.060
	p	1400.0	29.0	13.0	4.0	70.0		
3	h	0.9	0.9	6.0	54.0	?	621	1.830
	p	1650.0	10.0	45.0	14.0	39.0		
4	h	1.0	2.0	6.0	14.0	20.0	1000	2.340
	p	57.0	34.0	63.0	34.0	4.0		
					Layer V (h=?, p=63)			
5	h	1.0	1.0	9.0	39.0	?	543	2.120
	p	19.0	44.0	9.0	18.0	35.0		
6	h	1.1	3.3	8.8	87.0	?	817	0.470
	p	30.0	69.0	39.0	12.0	230.0		
7	h	1.0	4.0	12.0	34.0	?	985	1.900
	p	58.0	87.0	39.0	9.0	126.0		
8	h	2.0	6.0	16.0	50.0	?	848	1.654
	p	54.0	81.0	14.0	5.0	100.0		
9	h	1.0	3.0	13.5	17.0	?	1029	1.536
	p	100.0	150.0	22.0	14.0	43.0		
10	h	1.0	3.0	23.0	24.0	VH	815	1.629
	p	55.0	110.0	17.0	13.0	VH		
11	h	1.0	6.0	12.0	56.0	?	627	2.459
	p	80.0	53.0	9.0	11.0	78.0		
12	h	1.0	1.0	12.0	12.0	32.0	711	1.836
	p	38.0	19.0	32.0	21.0	5.0		
					LayerV (h=?, p=52)			
13	h	2.0	4.0	5.0	74.0	?	1133	1.870
	p	78.0	117.0	60.0	11.0	110.0		

(h in m.; p in Ohm-m; T in Ohm-m and S in Mho/m)

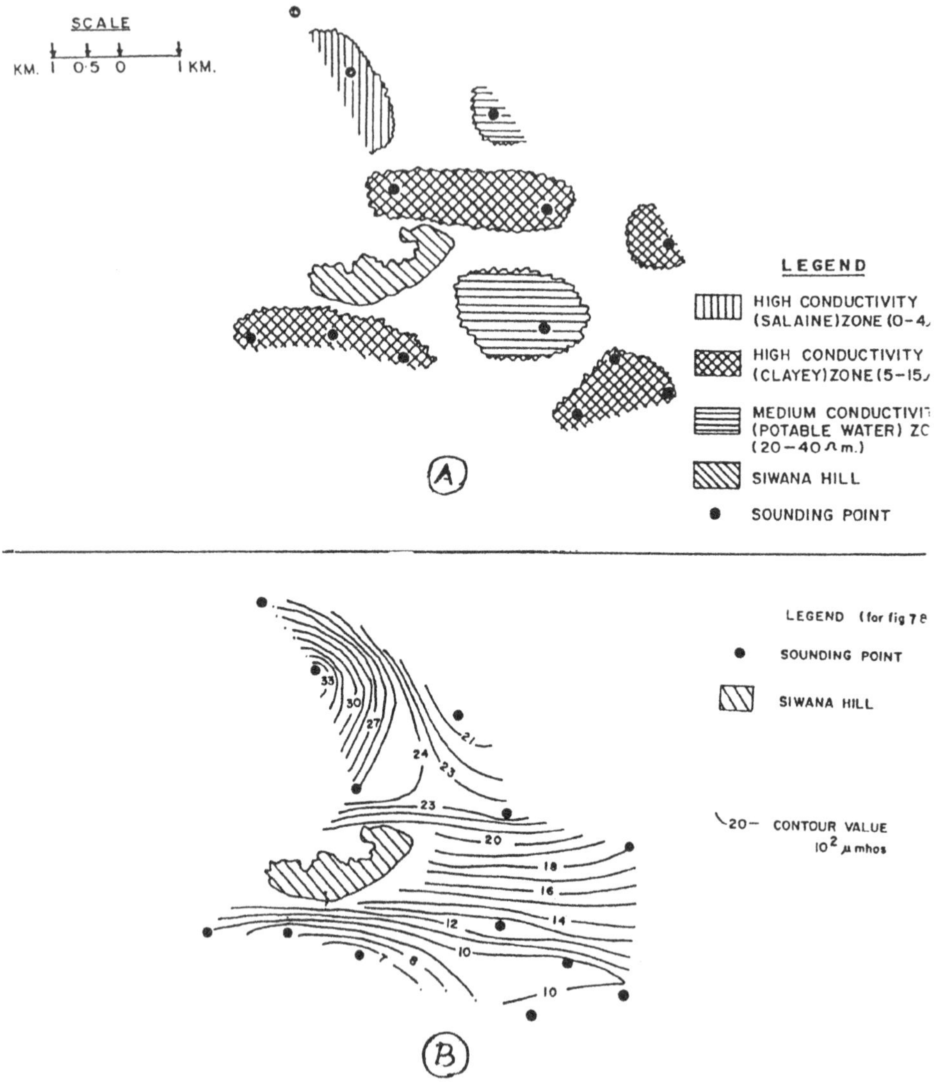

Fig. 2 (A) Water Poetential Zones Based on Resistivity and (B) Iso-electrical Contours of Ground Water Samples Around Siwana

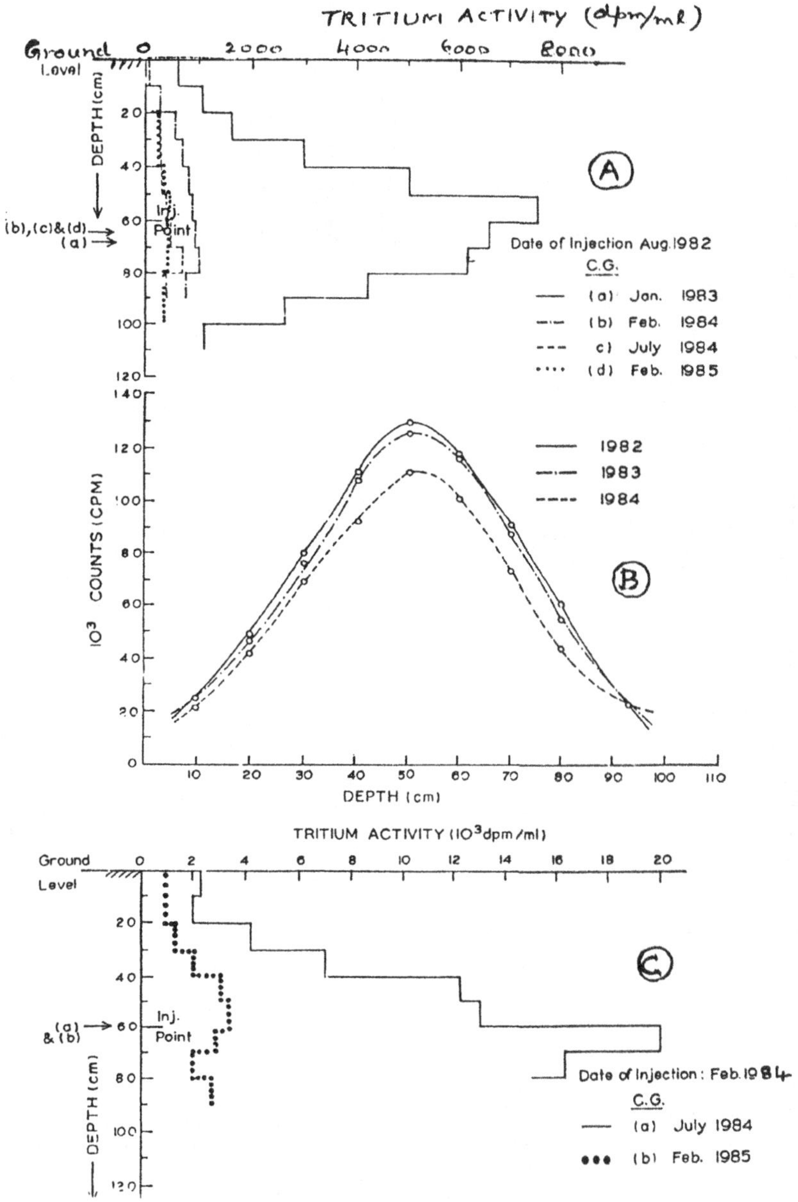

Fig. 3 Radio-tracer Profiles (A) and (C) Tritium in Siwana II & Siwana III and (B) Cobalt-60 in Siwana II Locations

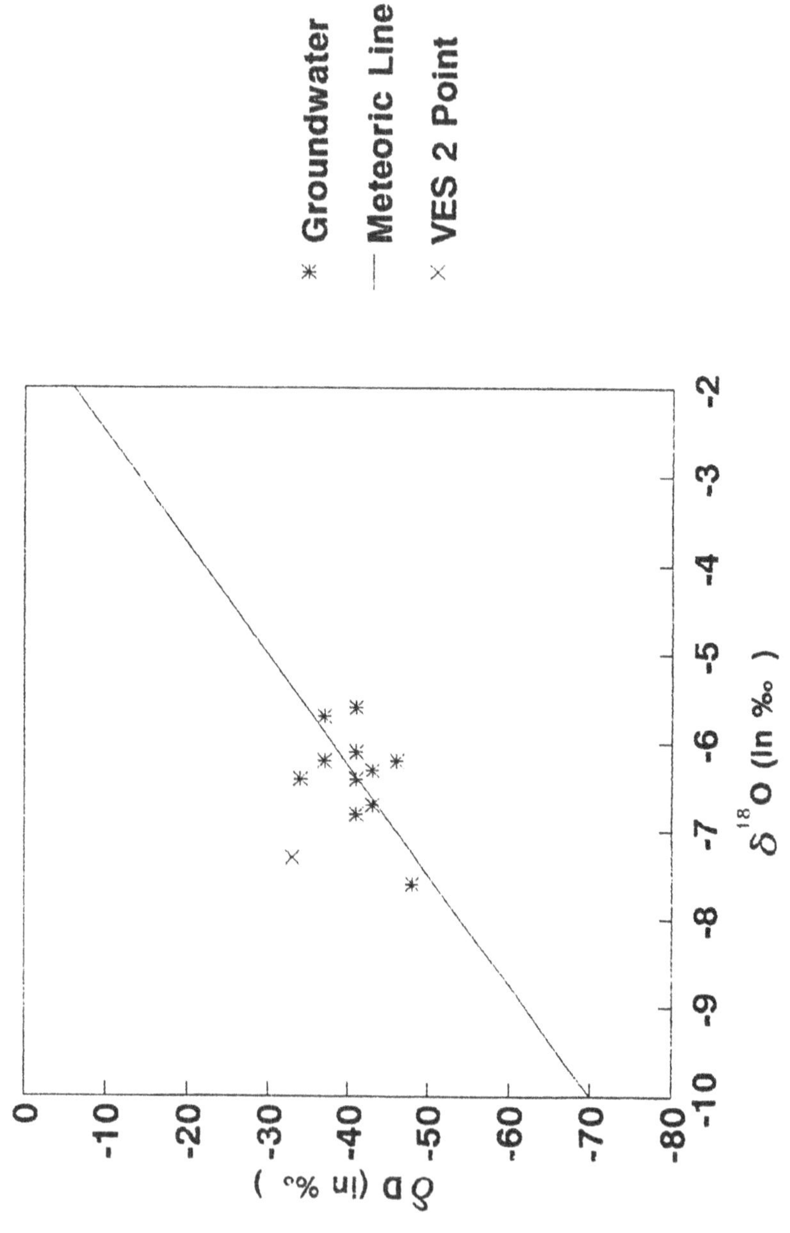

FIG 4 δD -- δ^{18}O PLOT OF GROUNDWATER SAMPLES FROM DEEP DUG WELLS AROUND SIWANA, RAJASTHAN

resistivity values ranges from 19 ohm-m to 100 ohm-m. Based on the iso-resistivity contours (Fig. 2 a) of the layer (representing aquifer) two types of groundwater potential zones; namely high conductivity zones (representing either clayey or saline formations) and medium conductivity zones (representing potable water) were identified. Iso-electrical conductivity (C) map of ground water samples (Fig 2b) shows the occurrence of relatively high salinity around VES 2 location and clayey zones south of Siwana. Henriet (1976) and Kelly (1977) have reported that the transmissivity and transverse unit resistance are linearly related. Accordingly, soundings at VES 2 and 5 are expected to have low transmissivities. Chandra and Athavale (1979) have reported that a combination of low S and high T values can be considered a potential aquifer in a region where the quality of groundwater is more or less uniform. In this area sounding 6 only has a combination of low S and high T and expected to have a potential aquifer relative to other points. The EC contours (Fig. 2b) also shows that the quality of ground water becomes more and more potable as we go towards Mawri (South of Siwana). But the quantity would be limited on account of clay particles encountered in the weathered zone.

Tritium and cobalt-60 tracer profiles (Figs 3 a, b and c) do not show any appreciable downward movement of the tracer at investigated sites inspite of 70.7 cm. rainfall during the observation period of about three years. The fractional recharge with this displacement works out to be about 1.5 per cent (Navada et al 1984). This very low recharge may be attributed to the clay particles encountered at varying depths in the area (Chandrasekharan, 1988).

In order to check whether there is any upward or lateral movement of tracer, soil samples at lateral distances of 20, 40 and 60 cm. on both sides of the central point at 60 cm. below ground level were collected from Siwana II location. Tritium counts--distance (Fig 5) indicate that the lateral distribution is more or less symmetrical following the well known diffusion equation

$$C\ (t) = C_o\ e^{-(r^2/4Dt)}$$

where c (t) is the tritium counts at distance r at time t,
c_o, the counts at the centre,
r, the lateral distance from the centre and
D, the diffusion constant.

It has also been observed that there is no upward movement

(Fig 2 a, b and c) expected during summer. This again confirms that the tracer is not lost and its movement is expected to move downwards if the local conditions favour.

Stable isotope contents of water samples, namely, δD and $\delta^{18}O$, show depleted values of heavier isotopes and have the values close to the weighted mean values of precipitation over a relatively humid region like Delhi. The δD--$\delta^{18}O$ diagram (Fig. 4) shows that the ground water samples have similar isotopic contents as those of world meteoric line which means that the ground water obtained at the investigated location is of meteoric origin and due to recharge occurred years back when the Thar desert was under relatively a more humid condition that at present. In the absence of Tritium and C-14 data of ground water samples from this region, it is not possible to substantiate the statement. However, most of the deep dug wells located adjoining to Siwana were found to have similar stable isotopic contents and do not have environmental tritium present (Navada et al., 1993). So if we agree for similar situation in this area also, then one can say that the ground water occurring here are of pluvial period and that the present day rainfall seldom recharges the groundwater. This is supported by both injected tritium and cobalt-60 tracer studies wherein we observed that there is very little movement of the tracer revealing only about 1.5 percent of recharge for over three years of experiments. In the light of these investigations, a general picture of ground water conditions in the investigated area has been obtained and presented in Table 2.

Conclusions

Geo-hydrological investigations reported above reveals that presence of clayey zones at varying depths are responsible for poor vertical recharge to the ground water around Siwana. The chemical quality of ground water is good at all places except for a small location north of Siwana. Stable isotope contents of the water sample reveal that the ground water in this area may either be recharged from the rainfall occurred years back when Thar desert was under a cool and humid condition or from a far off location wherein δ-values of rainfall are similar to those of a humid region.

Acknowledgements

Authors are thankful to Dr. S.M. Rao, Dr. S.V. Navada and Dr. Shivanna of BARC, for their valuable association in conducting isotope investigations. Grateful thanks are due to

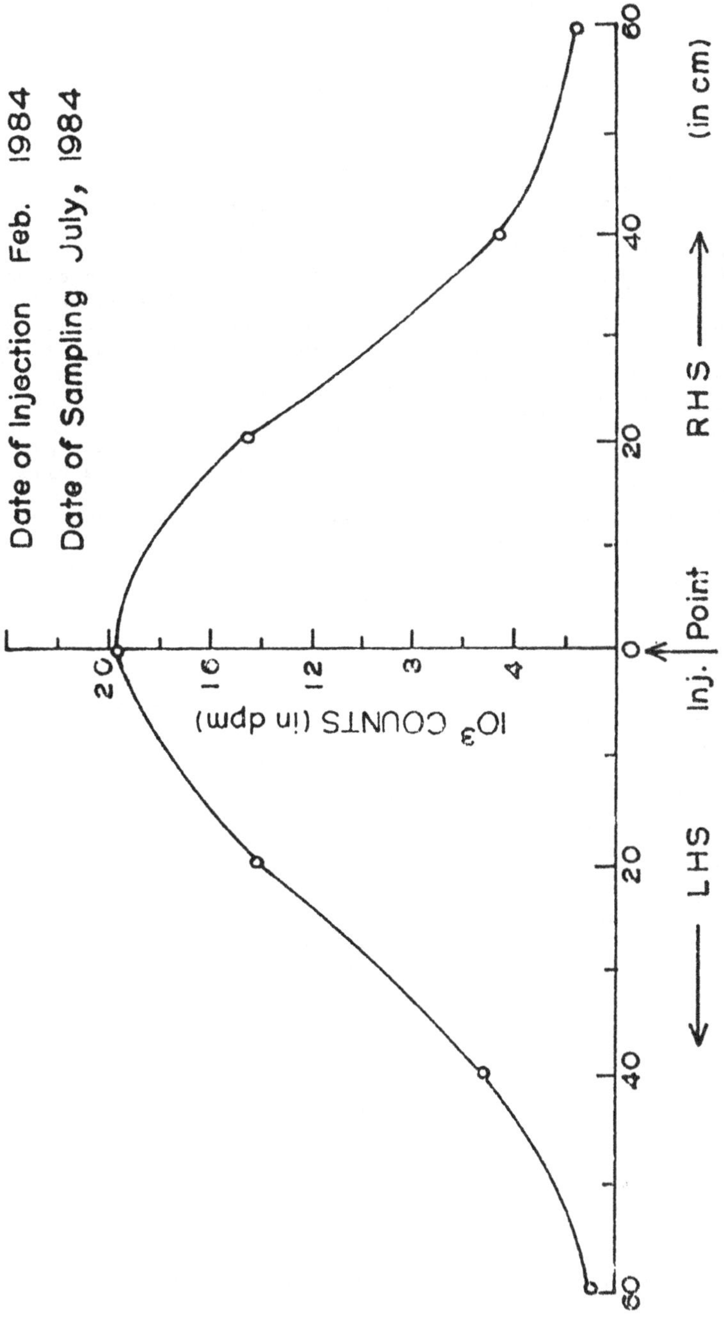

Fig. 5 Lateral Spread of Tritium Tracer Around the Injection Point at Siwana III Location

Table 2 Groundwater Conditions Around Siwana, Rajasthan

VES No.	Depth to Water (m.)	Depth to Weathered Zone (m.)	Depth to Bedrock (m.)	EC of Ground (umhos/cm)	Remarks
1	30	15	35	2700	Limited exploitation
2	25	15	?	3400	Limited exploitation
3	24	10	?	2500	Limited exploitation
4	27	22	70	1300	Well can be deepened
5	30	12	?	950	Limited exploitation
6	28	15	90	1050	Increase withdrawals
8	30	22	70	80	Increase withdrawals
9	27	20	?	1020	Limited exploitation
10	33	27	50	2350	Limited exploitation
11	22	20	45	1800	Limited exploitation
12	22	15	?	2000	Limited exploitation
13	25	12	?	600	Increase wiithdrawal

the Director, CAZRI, and Director, Defence Laboratory, Jodhpur for providing required facilities.

References

Adhyalkar, P.G. 1964 'Paleography, sedimentological framework and groundwater potentiality of arid zone of western Rajasthan', Proc. of the Symp. on problems of Indian Arid Zone, Jodhpur, pp. 4-17.

Chandra, P.C. and Athavle, R.N. 1979 'Close grid resistivity survey for demarcating the aquifer encountered in borewell at Koyyur in Lower Maner Basin', Tech. Rpt. No. GH. 11-GP-7, NGRI, Hyderabad, p.16.

Chandrasekharan, H. 1988 'Studies in ground water hydrology using geo-electrical and nuclear techniques' Ph.D. Thesis, Jodhpur University, Jodhpur, p. 224. (Unpublished).

Chatterjee, P.C. 1963 'Report on Geology, hydrogeology and utilization of Siwana development block region', Sectional Report, CAZRI, Jodhpur, p. 8.

Chatterjee, P.C. and Vengani, N.S. 1984 'Water balance studies in arid zone -- a case study', Indo-US Workshop on Arid Zone Research, Jodhpur, Jan, 9-14, 23.1--23.12.

Henriet, J.P. 1976 'Direct application of Dar Zarrouk parameters in ground water surveys', Geophys. Prospecting, 2 4, 2, pp. 344-353.

Kelley, W.E. 1977 'Geo-electrical sounding for estimating aquifer hydraulic conductivity', Groundwater, 1 5, 6, pp. 420-425.

Maillet, R. 1947 'The fundamental equations of electrical prospecting, Geophysics, 1 2, pp. 529-556.

Navada, S.V., Jain, S.K. and Rao, S.M. 1984 Symp. on Challenging Desert Problems, Defence Laboratory, Jodhpur (Unpublished)

Navada, S.V., Nair, A.R., Rao, S.M., Paliwal, B.L. and Doshi, C.S. 1993 'Ground water recharge studies in arid region of Jalore, Rajasthan, using isotope techniques, J. of Arid Environments, 2 4, pp. 125-133.

Rijikswaterstaat, 1969 'Standard Graphs for resistivity prospecting', European Association of Exploration Geophysicists, The Netherlands, p. 166.

SIGNIFICANCE OF PALAEOCHANNELS FOR HYDROGEOLOGICAL STUDIES - A CASE STUDY FROM ALLUVIAL PLAINS OF PUNJAB & HARYANA STATES, INDIA

K.P.Singh
Punjab State Council for
Science & Technology
SCO:2935-36, Sector 22-C,
Chandigarh-160022, India

ABSTRACT. The alluvial plains of Punjab and Haryana States, India have been characterised by the presence of Palaeochannels of Rivers Ghaggar, Sutlej and Beas. Such palaeochannels have been identified with the help of LANDSAT and LISS I,II data. These palaeochannels can be broadly grouped into two different types depending upon their climatic and sedimentary environments. As such two different models are proposed. In the first model, palaeochannels, when surrounded by comparatively finer sediments form water-logged and saline areas as free flow from palaeochannels is restricted. This results into development of saline water in the palaeochannels. In the second model, the surrounding material is either of same nature in terms of grain size, permeability etc or is having more permeability resulting in free flow of water from palaeochannels to the adjoining areas. In such cases palaeochannels are the source of fresh water specially in arid/semi-arid environments. In view of the above it is felt by the author that the significance of palaeochannels should be viewed/studied in relation to climatic and adjoining sedimentary environments as varied results are obtained. The models are supported by the field observations.

1.0 INTRODUCTION

The Indo-Gangetic alluvial plains are the richest source of ground water of our country. The plains are characterised by distinctive geomorphic features such as present day active channels, abondonded or burried channels, lakes, ponds, terraces, etc. A number of papers have been published on the geomorphic features of alluvial plains using aerial photographs and satellite data. Recently, it has been proposed that present day geomorphic features of the Gangetic plains are formed in response to climate fluctuations and fluvial adjustments during late Pleistocene - Holocene with a subdued role of tectonics of Himalaya (Singh 1987 and Singh et al. 1990). Further, Singh et al. 1990 on the basis of study of remotely sensed data opined that all the major rivers of Gangetic plain, many of them braided type today were of meandering in nature during last interglacial time (25 Ka). However, the applied aspects of such variations in grain size (sedimentary environment) on the ground water studies have not been sufficiently highlighted though generalised significance of palaeochannels has been stated by many workers (Singh et al.1980, Singh and Singh 1988; Gautam, 1990; Narsimhan, 1990 and others). In the present paper an attempt has been made to explain varied applications of such palaeochannels.

V. P. Singh and B. Kumar (eds.), Subsurface-Water Hydrology, 245–249.

2.0 STUDY AREA AND TECHNIQUES USED

The area forms parts of Punjab and Haryana States. Majority of palaeochannels, as report earlier by Yashpal et al. 1980, fall in the Districts of Ludhiana, Sangrur, Faridkot of Punjab State and Sirsa and Ambala districts of Haryana State (Fig.1). The area in most of the south-western parts face the acute problem of drinking water as ground water is largely saline and alkaline. They area is largely covered by older alluvium. Some areas around Ghaggar, Sutlej, Beas Rivers are characterized by newer alluvium near the river courses.

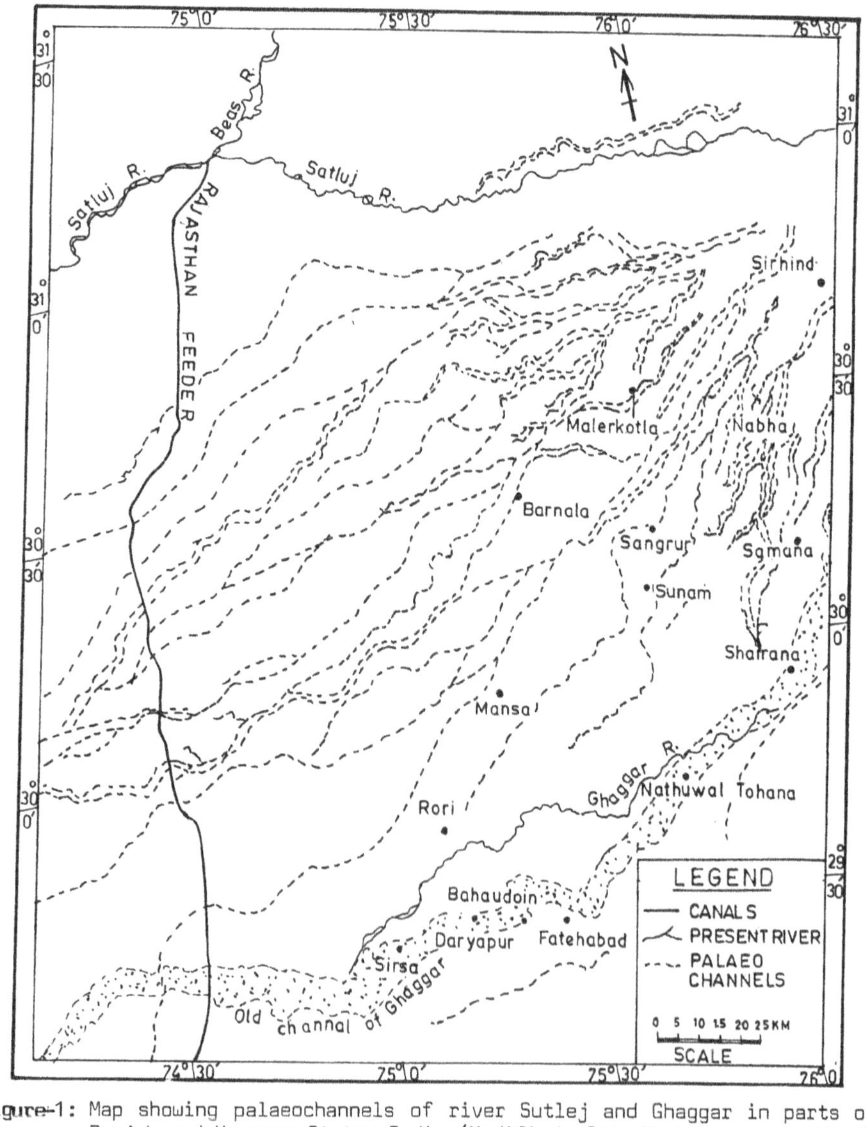

Figure-1: Map showing palaeochannels of river Sutlej and Ghaggar in parts of Punjab and Haryana States, India (Modified after Yashpal et al. 1980)

LANDSAT MSS data (band 5,7) TM data products, LISS-I, II (F.C.C.) have been studied on 1:1 million and 1:250,000 scale. Standard visual interpretation techniques were used. The author's observations and study support the findings of Yashpal et al.1990. Collateral data in the form of hydrogeological reports was used to study the bore hole data. Limited field checks have also been made for ground truth.

3.0 THE PALAEOCHANNELS

The palaeochannels which have been identified belong to Ghaggar, Sutlej and Beas Rivers. They mostly trend in north east - south west; north north east - south south west and east north east - west south west directions (Fig,.1). The palaeochannels are covered by 2-10m thick sheet washed clay/silt as found in few already drilled boreholes. The sediments of these palaeochannels comprise medium to coarse sands with few clay intercalations. At places the burried channels are marked on the ground by shallow linear depressions which are, however, imperceptible at most of the places. The palaeochannels are sinuous shaped with cut off meanders. The soils of some of these palaeochannels which are closed to present day rivers have comparatively higher moisture as compared to the surroundings due to accumulation of water in burried channels. The higher moisture of the soil cover of burried channels and the resulting vegetation concentration is picked up by satellite imagery and forms one of the basis of identification of burried channels. The length of major channels is variable and ranges between 10 and 30 kms and they are 100-300m wide. The sinuosity of channels does not show any marked trend except along the newer alluvium areas where cut off meanders, oxbow lakes etc. also occur as revealed by the study of aerial photographs (Singh and Singh, 1980).

4.0 PROPOSED MODELS AND SIGNIFICANCE IN HYDROGEOLOGICAL STUDIES

In the present paper two models are proposed. These models have different applications/implications in hydrogeological studies. Model 'A' (Fig.2, Case-I) shows a palaeochannel surrounded by finer sediments. The sediments in the palaeochannels are coarser and have comparatively larger permeability. If the permeability of surrounding sediments is K2 and of palaeochannels K1, then in this case K1>K2. This implies that water which may occur in the palaeochannels shall not be able to move freely in the adjoining environment as shown in Section-A A'. This results into locally rise in water levels in areas of palaeochannels, and if the water level is close to ground surface (within 2m) high rates of evaporation and capillary rise shall cause the development water-logged and saline soils. High evaporation rates in such areas may also impart local salinity to ground water as well. In the arid/semi-arid zones, such palaeochannels are associated with saline soils and poor quality of ground water. In case of second proposed model, the grainsize of adjoining sediments the palaeochannels is large and the permeability (K1) of palaeochannels is equal to or less than permeability of adjoining sediments i.e. K1=K2 or K1<K2. In such cases free flow of water from palaeochannels to the adjoining areas is possible/convenient as shown in Section-B B'. Such palaeochannels are normally characterised by good quality water, fit for drinking purposes. In the semi-arid zones, such palaeochannels form sweet water belts/zones which can be developed for drinking water supplies. Narasimhan (1990) has reported such old palaeochannels form productive aquifers. In Punjab and Haryana States bore hole data of Ludhiana and Sangrur districts have been studied. The borehole data supports the above proposed model.

Further, such palaeochannels where free flow of water takes place in surrounding environment offer also suitable sites for carrying out artificial recharge measures specially by the 'Basin method' in overexploiting areas.

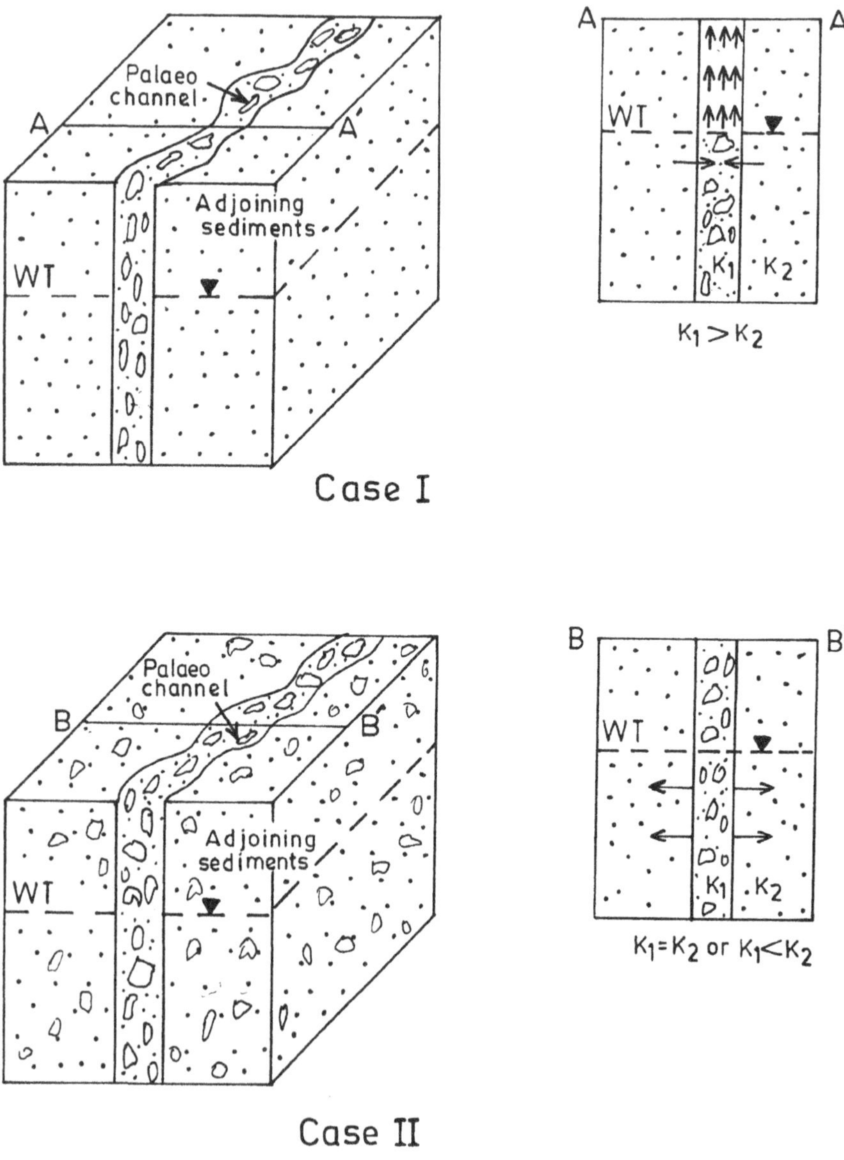

Figure-2: Figure showing two different models of palaeochannels. Case-I shows palaeochannels surrounded by finer sediments resulting in waterlogged areas and saline groundwater whereas in Case-II palaeochannels are surrounded by coarse grain sediments resulting into fresh water quality in palaeochannels.

5.0 CONCLUSIONS

Occurrence of palaeochannels in the alluvial plains of Punjab are reported on the basis of study of LANDSAT, LISS-I,II data. These palaeochannels can be broadly grouped into two different types, depending upon their climatic and sedimentary environments. In the first model, palaeochannels, when surrounded by comparatively finer sediments form water-logged and saline areas as free flow from palaeochannel is restricted. In the second model, the surrounding material is either of same nature in terms of grain size, permeability etc. or is having more permeability resulting in free flow of water from palaeochannels to the adjoining areas. In such cases, palaeochannels are the source of fresh water specially in arid/semi-arid environments. Varied applications/significance of palaeochannels is reported and the author feels that it is wrong to generalise the significance of palaeochannels in hydrogeological studies as is currently done. The significance of palaeochannels must be viewed/studied in relation to climatic and adjoining sedimentary environments.

6.0 REFERENCES

Chopra, S. (1990) 'A geological cum geomorphological framework of Haryana and adjoining areas for landuse appraisal using LANDSAT imagery'. Photonirvachak, **Vol.8**, No.1 & 2 pp.15-22.

Gautam, A.M. (1990) 'Application of IRS-IA data for delineating burried channels in southern part of Allahabad district of Uttar Pradesh'. Photonirvachak, **Vol.18**, No.3, pp.52-55.

Singh, I.B. (1987) 'Sedimentological history of Quaternary deposits in Gangetic plains. Indian Journ. Earth Sci., **V.14**,pp.272-282.

Singh, I.B., Bajpai, V.N., Kumar, A. and Singh, M. (1990) 'Changes in the channel characteristics of Ganga river during late Pleistocene - Holocene'. Jour. Geol. Sci. Ind. **V.36**, pp.67-73.

Singh, K.P., Tewari, B.S. and Williams, T.H.L. (1990) 'Application of LANDSAT imagery to ground water studies in parts of Punjab and Haryana States, India'. In Contribution of Space Observations to Water Resource Management, Pergamon Press Oxford, New York. pp.107-111.

Singh K.P. and Singh, B.D. (1980) 'Application of Remote Sensing Techniques to hydrogeological studies in parts of Ghaggar river basin, India'. Proc. 3rd Ind. Geol. Congr., Poona. pp.367-373.

Singh, K.P. and Singh B.D. (1988) 'Study of Hydromorphology of Doon Valley and Ganga-Yamuna Doab area using remotely sensed data'. Pub. Cent. Adv. Stud. Geol. Panjab Univ., Chandigarh, **Vol.3**, pp.77-82.

Narsimham, T.N. (1990) 'Palaeochannels of the Palar river west of Madras city : Possible implications for vertical movement'. Jour. Geol. Soc. **V.36**, No.5, pp. 471-474.

Yash Pal, Sahai, B., Sood, R.K. and Agarwal, D.P. (1980) Remote Sensing of the lost Saraswati River. Proceedings of the Indian Academy of Science (Earth and Planetary Sciences). **Vol.89**. No.3, pp.317-331.

Integrated use of remote sensing and GIS methods for ground water exploration in parts of Lalitpur district, U.P.

A.K. Saraf and S.K.Jain*

Department of Earth Sciences
University Of Roorkee
ROORKEE - 247 667.

* National Institute Of Hydrology,
ROORKEE - 247 667.

ABSTRACT

Remote sensing provides much of the information that is input to a GIS from global scale vegetation and climatic data to the ground water exploration. The two technologies provide complementary capabilities. Remote sensing analyses are improved by the verification data retrieved from the GIS, and application of GIS can greatly benefit from the information that remote sensing data contain. Often the image data are the most current spatial information available for an area. The use of digital image data offers the additional advantage of a computer compatible format that can be input directly to a GIS.

The present study demonstrates that the integrated use of remote sensing and GIS methods in ground water exploration not only improves the quality of geographic information but also enable information previously unavailable to be economically produced. In the present study various sets of information layers such as geological, soil,topographical and groundwater information have been applied along with IRS-1A LISS-I data for ground water exploration for Shahjad catchment located in Lalitpur district. Groundwater recharge map has been produced with the help of groundwater level data and specific yield information of different rock formations.

INTRODUCTION

Groundwater constitutes an important source of water supply for various purposes like domestic needs, supply for industries and for agriculture etc. The term groundwater is usually referred for the subsurface water, that occurs beneath the water table in soils and geological formations called aquifers. Frequent failure of the monsoons and unassured supply of surface water has resulted in extensive search for groundwater resources. The occurrence and movement of groundwater in any terrain is mainly controlled by geology, topography, landuse, soil, geomorphology and structures. Depending on the terrain characteristics, one or several of these parameters contribute to the localization of groundwater.

V. P. Singh and B. Kumar (eds.), Subsurface-Water Hydrology, 251-259.

The conventional approaches for groundwater investigation are ground based surveys and exploratory drilling which are time consuming and uneconomical. The remote sensing technique provides timely information on geological and hydrological parameters required in groundwater hydrology. Repetitive coverage of the same area of the earth provides temporal and real time information on static and dynamic resources. The geological and hydrologic information can be inferred by analysis and interpretation of satellite data.

The exploration and exploitation of this subsurface resource depends upon the understanding of the expected sub-surface hydrological conditions through surface expressions. The different types of digital enhancement techniques using remotely sensed data help in the extraction of specific features that act as ground water indicators and ultimately lead to the preparation of maps indicating groundwater prospective zones. The features which can be extracted from satellite data include landforms, drainage patterns,landuse/landcover,soils,linear and curvilinear features.

The accuracy of classification of groundwater potential zones can be improved by using the remotely sensed data along with ancillary data like topographical, meteorological and geophysical data. An effective and integrated utilization of remote sensing and other data requires an efficient spatial data processing system which can handle both spatial and non spatial data. A Geographic Information System (GIS) is rapidly evolving technology that consist of computer based programs containing specialized algorithms and associated data base management structures, frequently in an integrated package. These systems re designed to store information about the location, topology and topography of spatially referenced objects. A GIS consists of four basic components, data gathering and input, geographic data bases, data analysis and modelling and data visualization and presentation. Data in GIS have both locational and nonlocational (i.e. attribute) characteristics. GIS deals with data related to location in real world space called as geographic or spatial data. Spatial data base is a collection of information about things and their relationships to each other. GIS technology can make valuable inputs as it provides enormous potential for effective storing, handling, manipulating and analysing multiple spatial data sets in a single analysis at a high speed, unmatched by any other methods.

The Shahjad catchment a part of Lalitpur district located in Bundelkhand region has been selected for this study. The area lying between latitudes 22 30 to 25 30 North and longitudes 78 15 to 78 45 (Fig.1) consists of agricultural land with local undulations. The area falls in the survey of India toposheets No. 54 L. Geologically, the area comprises crystalline rocks of Bundelkhand granite , Deccan trap and Vindhayan formations. The drainage pattern in the area is dendric. The area experiences subtropical climate with very hot summer and mild winter. The

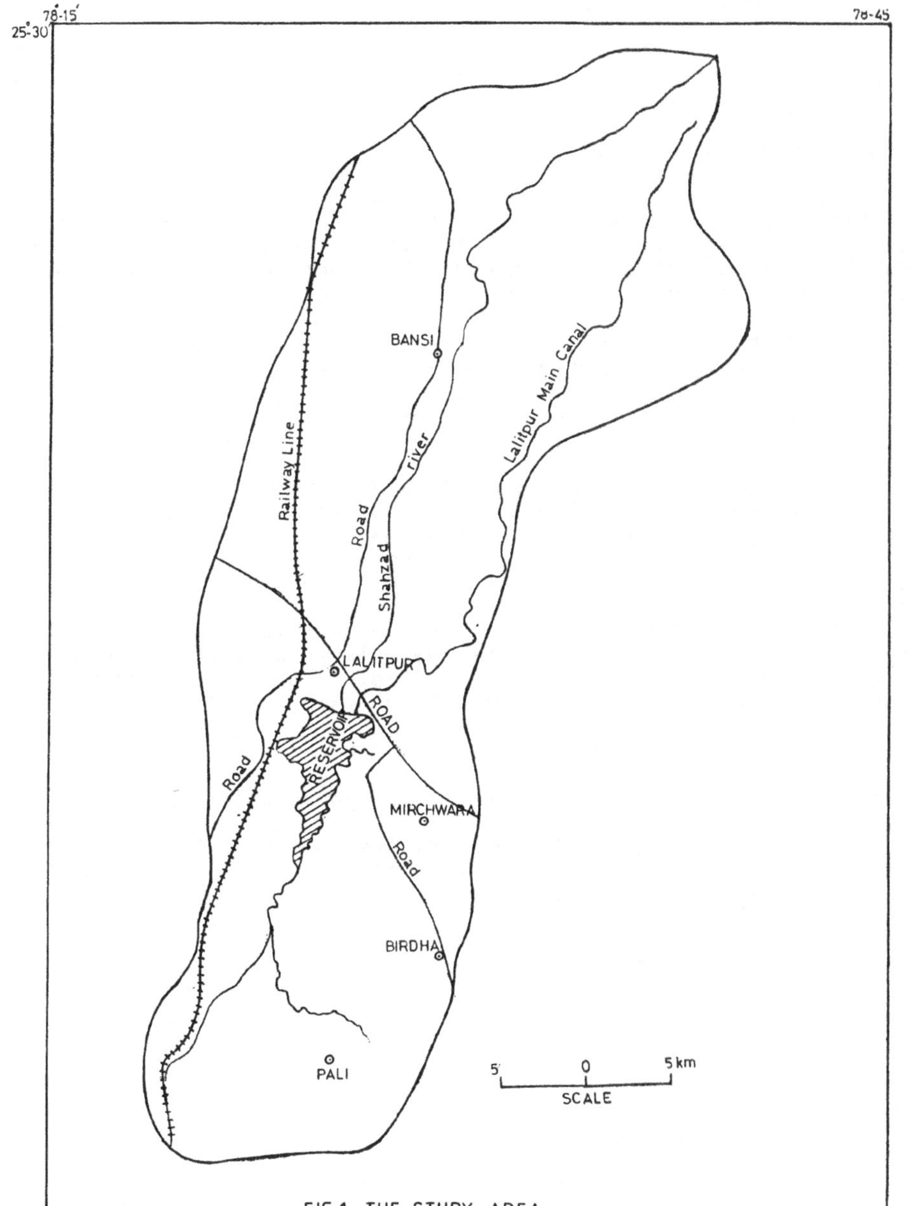

FIG.1 THE STUDY AREA

average rainfall is about 900 mm.

In addition to the recharge from rainfall, a significant source of recharge to groundwater basin is through the rivers and reservoirs located in the area.

DATA USED AND METHODOLOGY

The GIS used to carry out the present work is ILWIS (Integrated Land and Water Information System). This GIS has been developed at International Institute of Aerospace Survey and Earth Sciences (ITC), Enschede, The Netherlands. The main characteristics of the system are :

- PC based system
- A tabular, vector and raster data base integrated in one system
- Image processing capabilities, fully integrated into the GIS.

From the IRS-1A LISS I satellite data which was available on CCT (path/row no. 27-51 and date 28.5.89),a part covering the study area was extracted. The base map of the study area was prepared using Survey of India toposheets no. 54 L. This map was digitized using DIGITIZE module of the ILWIS. Geometric transformation are used to assign ground coordinates to a map or data layer within the GIS to adjust one data layer so that it can be correctly overlaid on another data layer of the same area. A green vegetation index (GVI) map (Fig. 2) using all the four bands was prepared with the help of following relation :

$$GVI = [(B4-B3)/(B4+B3)*127]-128$$

127 and 128 are scaling factors to avoid negatives.

Soil map (Fig.3) of the study area was prepared with the help of existing map and topographical map. In this map the different symbols represents the soil types and characteristics as given below.

s1	Skeletal, lithic ustorthents	deep to very deep soil that have a shallow horizon in which carbonate have accumulated.
G1	Skeletal, lithic ustorthents	-do-
G21	Skeletal, lithic ustorchrepts and typic ustorchrepts.	-do-
G42	Typic ustorchrepts	-do-
G22	Skeletal, vertic haplustalfs	These soils have clayey texture and have montmor--illonitic type of clay
G41	Typic chromsterts and udichaplustalfs	Very dark brown to very dark greyish brown

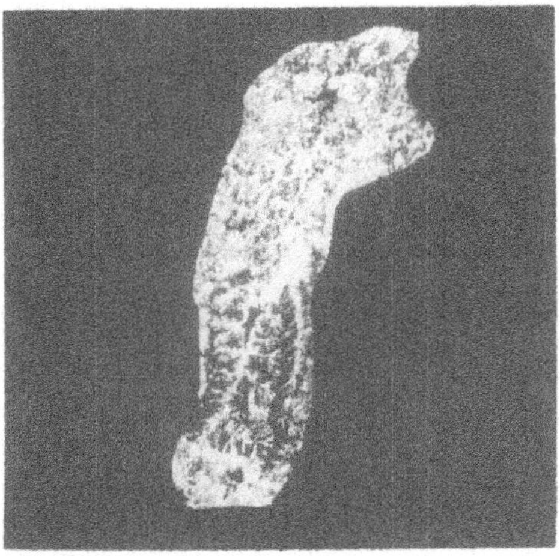

FIG.2 : Green vegetation index map of Sahjad catchment

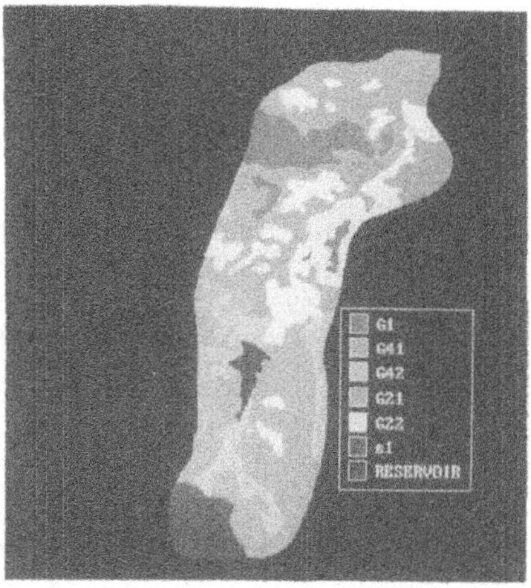

FIG.3 : Soil map of Shahjad catchment

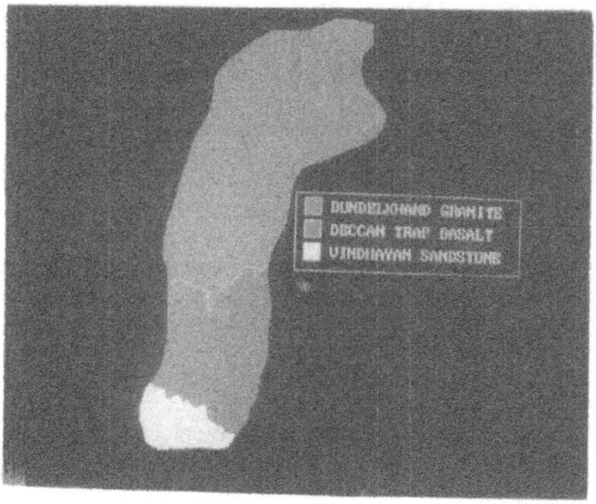

FIG.4: Geological map of Shahjad catchment

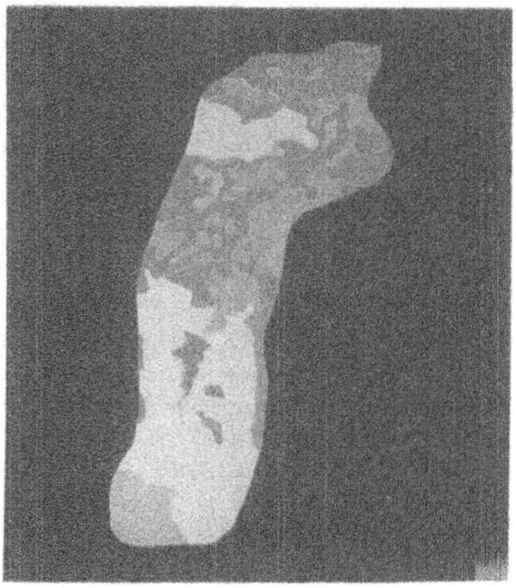

FIG.5: Drainage map overlayed on soil map of Shahjad catchment

Drainage map was prepared from topographical map. These maps of drainage and soil were digitized in ILWIS. The area mainly consists of three major geological units (Fig. 4) are Granite, Deccan tarp and Vindhayan sandstone. These three units were extracted from the satellite image with the help of digitization on the image.

Average pre and post monsoon data of ten years of water level (1979-1984) was available for the following six stations.

WATER LEVEL (m)

Station	Pre monsoon	Post monsoon
Birdha	8.36	3.31
Mirchwara	5.24	3.07
Pali	6.69	3.71
Bansi	11.71	5.37
Lalitpur	3.23	1.58
Gangorali	6.41	2.72

By linear interpolation in GIS of above six well data of pre monsoon and post monsoon, depth to groundwater level maps were generated.

ANALYSIS AND DISCUSSIONS:

The following maps have been prepared using satellite data/digitization of existing maps: Soil type map, Drainage pattern map, Geological map, green vegetation index map and depth to water level map. Using these maps the analysis of identification of ground water potential zones have been carried out.

The study area is underlain by three distinct hardrock formations viz granite, deccan trap and vindhayan sandstone. The granite are of Archean age and covers topmost area of the catchment. The deccan trap are of upper cretaccous and occupy middle portion of the catchment. The Vindhayan sandstone are of pre cambrian age and covers the lower part of the catchment. These are generally hard ,compact and quartzitic with infrequent jointing. Groundwater movement is mainly through joints and fractures. Granite in general is poor to recharge but the study area has enough lineaments in granite so recharge is high.

The soils of a region play a vital role in understanding the groundwater recharge. The black soil terrain which is in the middle part of the catchment is a poor recharge zone. Rest of the area in general has good recharge. Fig. 5 shows the map in which drainage map is overlayed on soil map, which shows a close relationship between different soil type with drainage network and its density.

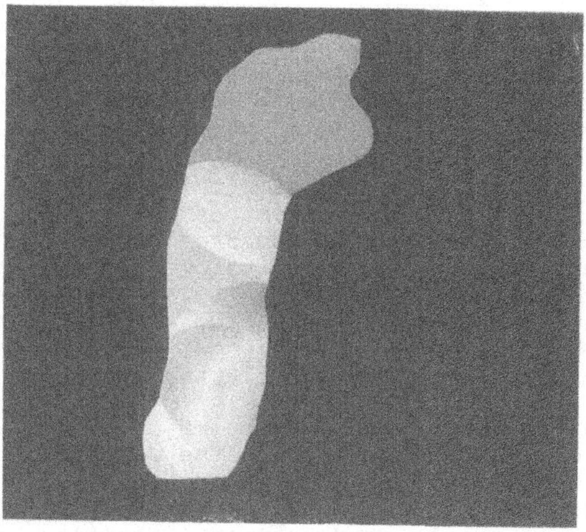

FIG.6: Photograph showing colour coded water level fluctuation image. Reddish colour shows maximum fluctuation whereas yellowish colour shows medium and bluish colour shows minimum water level fluctuation.

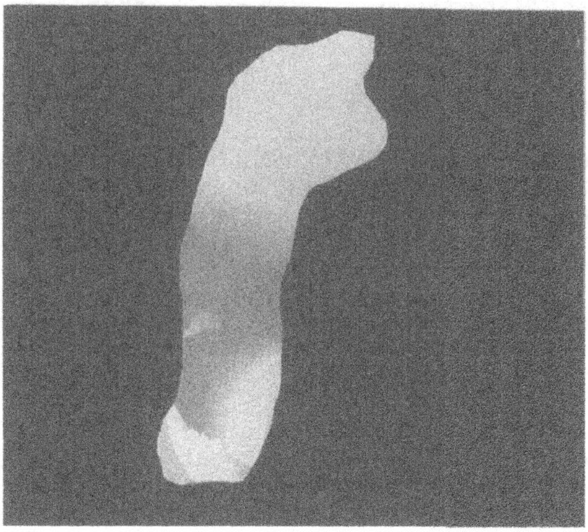

FIG.7: Photograph showing colour coded groundwater recharge image. Yellowish colour shows high groundwater recharge and bluish colour shows low groundwater recharge zones.

In the green vegetation index map(after stretching) water can be seen as black and green vegetation light. As seen from the fig. 2 the upper and lower part of the catchment is light so presence of vegetation represents the availability of more groundwater in this part.

For a proper evaluation ,development and management of groundwater resources of an area ,it is essential to study the changes in water level fluctuations of the groundwater regime.To study the water level rise from premonsoon to postmonsoon a fluctuation map has been prepared using interpolation from point data of six stations (fig 6). In this map upper portion in granite zone fluctuation is maximum which is represented by reddish colour, yellowish colour shows medium fluctuation while bluish colour represents minimum fluctuation.

Recharge to the groundwater mainly occurs as vertical percolation of rain water through the weathered and surfacial mental or through open fractures in the outcrop area. Some recharge also takes place through induced infiltration from reservoir,streams,canals and applied irrigation water. Most of the recharge in the area occurs during the SW monsoon period i.e. June to September. The recharge to groundwater has been estimated estimated by the analysis of water level fluctuations for the presents study. For this purpose the specifies yield for three geological units are taken as following (Karanth 1987).

Type of Geological Unit	Range	Specific Yield Value
Granite	0.5-0.4	2.2
Deccan trap	0.2-16.0	3.0
Vindhayan Sandstone	4.0-8.0	6.0

The Recharge image (Fig.7) can be generated by multiplying specific yield with water fluctuation maps. In this colour coded image of groundwater recharge yellowish colour areas show high recharge while bluish colour area show low recharge. In the upper part of the catchment , data of water level is only available for one station (Birdha). Therefore, variation in waterlevel fluctuation could not be interpolated. Only a uniform value is available regarding water level fluctuation.

CONCLUSION:

Groundwater is very important national resources and is widely used for different purposes like drinking,irrigation etc. The study reveals that integration of geological, soil, drainage and water level fluctuation data provides better understanding of groundwater resources. If good amount of data is available then integration of data in GIS environment for groundwater studies can be done more efficiently.

REFERENCES :

Agarwal, C.S. (1988),' Hydrogeological investigation in Bundelkhand region using remote sensing techniques', M.E. dissertation, Dept. of civil Engg., University of Roorkee, Roorkee.

Chaturvedi, R.S., D.C. Bhatacharya, P. Kamal, J. Krishnamurthy and N. Sunder Ramam, (1983) " Integrating remote sensing techniques in groundwater exploration - A typical case study from Bundelkhand region in U.P.," Proc. Nat. Symp. on Remote Sensing in development and management of water resources, Ahemadabad, Sep. 25-27,1983.

Karanth K.R.(1987) ,'Groundwater development and management ', McGraw Hill publishing company limited, 1987.

Krishnamurthy, J. and P. Manavalan (1992)," Application of digital enhancement techniques for ground water exploration in a hard rock terrain, I J R S Vol. 13, No. 15 ,1992.

Mishra, D., G.L. Rao, H.Singh and A. Jeyram (1986)", Some results of groundwater targetting in Granitic terrain of Lalitpur district, U.P., Jour, Assoc. Expl. Geophys. Vol. VII, No. 2, April 1986.

Rao ,M.B.S., A.L Haldar and Rajeev Mohan, (1992)", Groundwater targetting in Lalitpur district, U.P. through digital image processing, - A case study, Proc. Nat. Symp. on remote sensing for sustainable development, Nov. 17-19, Lucknow, 1992.

Srivastwa N.K. (1993),' Technique of processing and interpretating groundwater data using GIS'. M.Tech. dissertation in Applied Geology, U.O.R , Roorkee.

REGIONAL GROUND WATER MODELLING USING FINITE ELEMENT METHOD - A CASE STUDY

V. VENKATESHWAR RAO[1] AND P.B.S. SARMA[2]

1 Scientist, Water Resources Division, IIRS, Dehradun-248001
2 Professor, Water Technology Centre, IARI, New Delhi-12

ABSTRACT

Judicious management of the water resources of a groundwater basin needs a comprehensive understanding of the total system and its response to recharge and various pumpages contemplated. The groundwater basin of D-53 major distributary command area, SriRamSagar project, Andhra Pradesh was simulated using the Finite Element numerical method. Net recharge to the aquifer was assessed by estimating various components of the water balance by adopting the norms of Central Ground Water board estimates Committee with suitable modifications based on the soils, crops and climate of the area under consideration. The approach using the groundwater level fluctuation as an indicator of spatial distribution of net recharge was adopted. Sensitivity analysis was performed to investigate the models sensitivity to its parameters viz. Hydraulic conductivity and storage coefficient. The model was validated against the historical observed data of groundwater levels in the study area and found a close agreement between observed and model computed Ground Water levels for three years from 1983-84 to 1985-86. Since the groundwater levels were rising in the study area, the model was used to determine the stabilizing pumping and recharge rates under a proposed crop-water management scheme.

INTRODUCTION

Intensive agriculture based on high yielding crop varieties, has exacting demands on the inputs, of which water is the major one. To get the optimum benefit per unit of water used, it is essential that water must be used efficiently. Efficient water management includes judicious use of both surface water and groundwater along with other inputs, and proper manipulation of agronomical practices for maximizing the crop production. One aspect of water resources utilization in major irrigation projects which has not received adequate attention in water management is the increasing potential for groundwater development in canal irrigated areas. In several canal commands of the country, excessive use of surface water, without giving sufficient attention to the groundwater has resulted in rising water table causing water logging and salinity conditions and turning part of the fertile land into barren area. Infact, the potential for groundwater development has increased because of recharge to aquifers from water conveyance and distribution systems of the irrigation canal networks. (Water Technology Centre, 1983). Conjunctive use of groundwater and the surface

V. P. Singh and B. Kumar (eds.), Subsurface-Water Hydrology, 261–275.

water resources is of utmost importance to prevent this problem. This fact is duly recognised in India and constitutes one of the strategies for sustainable development of irrigated agriculture (Ministry of Irrigation, 1984, Planning Commission, Government of India, 1985).

An adequate estimate of the availability of groundwater storage in a basin requires determination of the groundwater basin boundaries, aquifer dimensions, aquifer characteristics and water budgeting. Water budgeting helps to evaluate the net available water resources both surface and subsurface, and to assess the impact of existing water utilization pattern and practices. The need for long and tedious calculations have often precluded quantitative description and analysis of groundwater conditions over a basin. However, the advent of Computers and the concomitant developments in numerical models helped better understanding of the behaviour of the system. In particular, the groundwater basin simulation models help obtain the information needed in terms of the response of the basin to different levels of pumping and recharge. This information constitutes the scientific basis for development of guide lines for planning sustainable use of groundwater resources in command areas.

The specific objectives of this study are to develop a digital simulation model of the groundwater basin of a canal command area and then to use the model to asses the spatial distribution of groundwater potential to be used to develop conjunctive water use management plans for an increased agricultural productivity in a sustainable manner.

STUDY AREA

The present study is carried out with respect to the command area of D-53 major distributary of the SriRamSagar major irrigation project (SRSP), Stage-I, Andhra Pradesh. The D-53 distributary is the largest distributary in SriRamSagar Project. The gross command area of the distributary is 833.18 sq.km., spread over six mandals (administrative units) of Karimnagar district, bounded by Godavari river in the north, Peddavagu-II in the west, Peddavagu-III on its east and south east and SRSP main canal which occupies in part the south of the study area (Fig. 1). The climate of the area is semi arid with dry hot summers and mild winters with an average annual rainfall of 944 mm. The rainfall in the study area is seasonal, with about two-thirds of annual rainfall being received in the monsoon season (June-October). Two-thirds of the study area consists of coarse textured soils (Red soils) and the remaining one-third is under clay soils, range mostly from shallow to moderately deep. The general cropping pattern is single wet and single I.D. (Irrigated Dry) crop in the ratio of 1:2 during Kharif and Rabi season respectively. The major crops grown are paddy, maize, jowar, groundnut, chillies and pulses etc.

IRRIGATION SYSTEM

Sources of irrigation in the study area are canals, tanks and

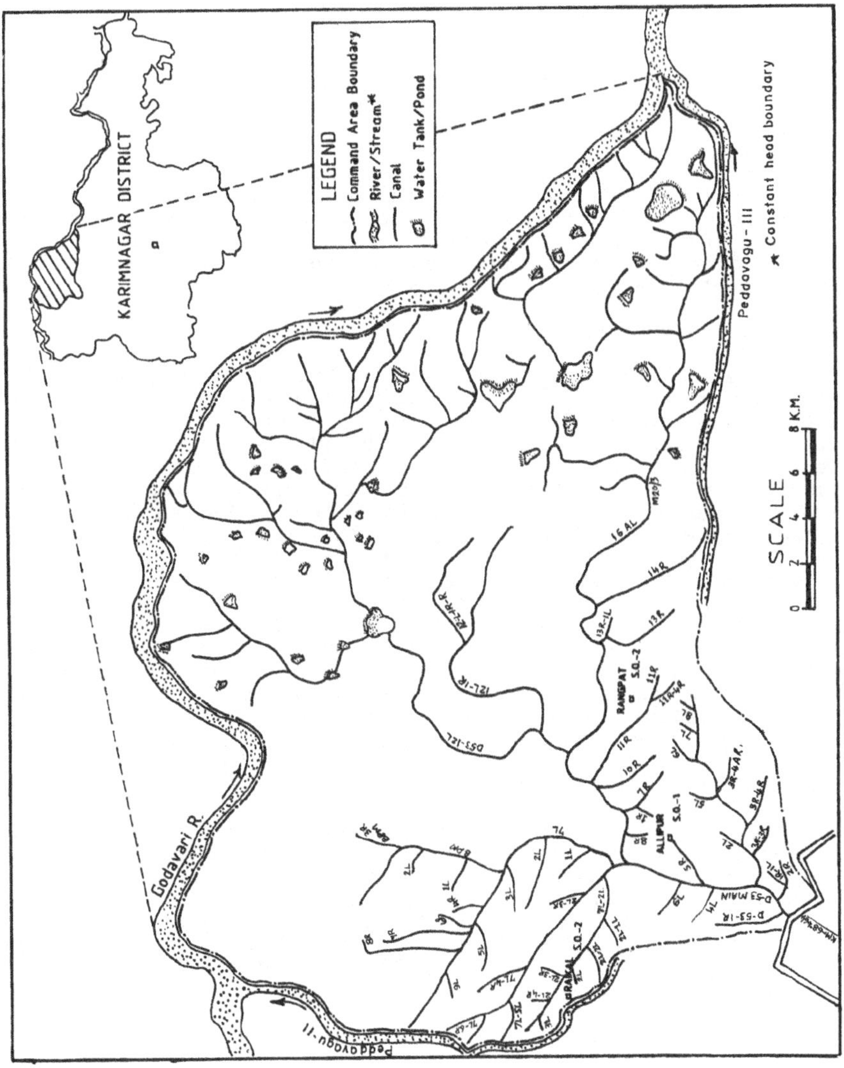

Fig. 1. LOCATION MAP OF STUDY AREA

wells (mostly dug wells) and cover 50, 23 and 27 percent of the net area irrigated respectively (APSGWB, 1984). The D-53 distributary traverses a length of 53.1 km. with a designed discharge capacity at the off take point of 22.60 cumecs (798 cusecs) commanding 25233 ha. of which 7349 ha. is localised as wet and 17884 ha. as irrigated dry. All the minor distributaries are unlined. Canal net work (Fig 1) of D-53 major consists of 381 outlet pipes which have been earmarked separately for wet and I.D. command areas as wet pipes and I.D. pipes. In the north eastern part of the command, the minor distributaries are connected to the reservoir/tanks and the canal network is extended further below the tanks, most of which serve the localised wet areas.

OCCURRENCE OF GROUNDWATER

The command area is underlain by Archean group of rocks consisting predominantly of granites which are medium to coarse grained, grey and pink in colour. Groundwater in these geological formations occur under watertable condition in the weathered zone, joints, fissures and fractures beyond the weathered zone. Weathered rock aquifers are developed through dugwells, generally rectangles in shape. The thickness of weathered zone is found to vary from 12 to 19 mtrs. The yield of wells range from 40 to 150 m^2/day, transmissivity of aquifer zone ranges from 24 to 96m^2/day, storage coefficient varies from 0.001 to 0.019, the specific capacity of wells ranges from 0.0106 to 0.159 m^3/minute/mtrs draw down and the wells sustain continued pumping for 3 to 4 hours daily (APSGWB, 1984). The deeper weathered and fractured rock aquifers are developed by drilling extension bore at the bottom of open dug wells. The quality of groundwater falls in`good to permissible' and `excellent to good' categories for domestic and irrigation purposes (Veeranna, 1990).

MONITORING OF OBSERVATION WELLS

There are 78 observation wells which are a part of the net work of observation wells in SRSP in the study area are being monitored by State Groundwater Department since 1972. A study conducted (Bowonder and Ravi, 1983) indicates that water table levels have been rising since 1970, the year in which the irrigation system was commissioned in SRSP. An a result of this rising water levels, 32 percent of the SRSP command area already been water logged by the end of 1982. The study suggests the alarming rate of groundwater rise is to be arrested with an immediate effect and calls for conjunctive water use planning in the area.

GROUNDWATER BASIN SIMULATION MODEL

The prerequisite in development of a groundwater basin model is the mathematical formulation of the physical process of the groundwater flow through aquifer. The equation of a two dimensional transient flow through porous matrix, which is considered in this study, is given by

$$\frac{\delta}{\delta x}\left[T\frac{\delta h}{\delta x}\right]+\frac{\delta}{\delta y}\left[T\frac{\delta h}{\delta y}\right]+N = S\frac{\delta h}{\delta t} \qquad (1)$$

where: T = T(x, y), transmissivity of the aquifer = kb, m^2/sec; b = average thickness of the aquifer, m; k = hydraulic conductivity, m/sec.; S = S (x, y), storage coefficient of the aquifer; h = h (x, y, t), hydraulic head in the aquifer, m; N = N (x, y, t), net recharge per unit area, $m^3/sec/m^2$; t = time; and x, y are Cartesian co-ordinates in horizontal plane.

In this study, Finite Element digital simulation model proposed by Swain (1978) was used with appropriate modifications to solve two-dimensional transient-state problem of groundwater flow in an unconfined aquifer of D-53 major command of SRSP. The program utilizes triangular elements and linear trial functions and allows specified-flux and specified-head boundaries. Sources and sinks are allowed at nodal points, and flow rates may vary from time to time.

DEFINITION OF FLOW DOMAIN

Total geographical area was divided into 158 triangular elements with 107 nodal points so as to produce minimum half-band width of 20. Size of the triangles have been adjusted to conform to the shape of the boundary and size of the elements were reduced near the boundaries to enable better simulation (Fig. 2).

Due to inadequate data on aquifer tests and well logs, and other geohydrologic information pertaining to the groundwater system, it was difficult to obtain spatial distribution of the aquifer parameters namely, transmissivity and storage coefficient. The earlier studies conducted (Baig and Prabhakar, 1980; Srineevasulu et.al., 1984; and Murthy, 1987) in the adjacent groundwater basin which has similar geohydrological formations by considering it as a homogeneous aquifer. Since most of the study area is underlain by pink granites, aquifer of the study area was assumed to have an average weathered zone thickness of 15.5m, average transmissivity value of 79.5 m^2/day and average storage coefficient value of 0.01.

INITIAL AND BOUNDARY CONDITIONS

The time period considered in this study is 1983-84 to 1985-86. As the first step, the model was used to simulate the groundwater levels in the year 1983-84. Hence, the time period considered in the model was one year divided into 12 time steps, each consisting of 30 days (a month).

Initial conditions are the specification of the values of the hydraulic heads over the flow domain at the start of the time domain (1983-84 to 1985-86). Hence, the observed water table contours in the month of April, 1983 were considered to describe the initial conditions.

River and streamlets constitute natural constant-head boundaries. The general nature and extent of the aquifer boundaries are shown in Fig. 2. The remaining portion of the boundary which

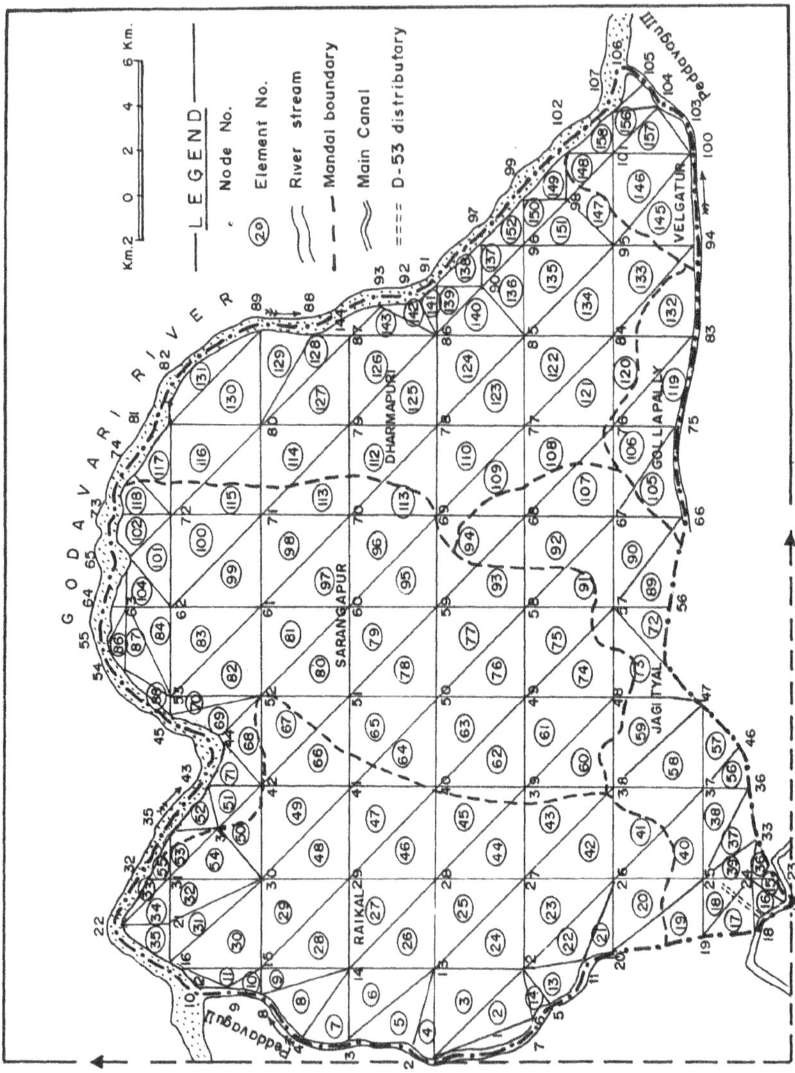

Fig. 2. DISCRETIZED PLAN OF D-53 COMMAND AREA

is not a natural boundary. However, the water levels in the month of April for the period 1983 to 1986 in observation wells near this portion of the boundary and the nodes located at this segment indicate that they are more or less constant. Hence, this portion of the boundary was also designated as a constant head boundary.

SOURCES AND SINKS

Sources and Sinks specify the recharge and discharge respectively for the groundwater system over the time domain to be simulated.

Pumping rates : Groundwater in the basin is being developed by dug wells and dug-cum bore wells. Inventory of observation wells for each of the six mandals was obtained (APSGWD, 1984-86 and CPO, 1984-86). Groundwater draft for each mandal was estimated by adopting 0.0028 MCM (million cubic meters) and 0.0218 MCM of average annual draft for dug wells and dug wells fitted with pumps, respectively. Since, the spatial distribution of pumpage, in each mandal was not known, it was uniformly distributed within the mandal. Since exact time distribution of the pumpage (total draft) was not known, the total draft was distributed uniformly throughout the year, i.e., from April to next March.

Assessment of net regional groundwater recharge: To estimate the total groundwater potential available, it is essential to know net regional recharge to groundwater, which was estimated as below :

$$R_n = R_g \pm Q_g - Q_p \qquad (2)$$

where: R_g = gross recharge to groundwater; Q_g = groundwater inflow/outflow into or from the region; and Q_p = groundwater draft.

The gross recharge (R_g) to groundwater was estimated by using the following equation :

$$R_g = R_e + R_c + R_d + R_t + R_{ci} + R_{wi} + R_{ti} \qquad (3)$$

where: R_e = recharge from rainfall; R_c = recharge due to seepage from major distributary; R_d = recharge due to seepage from minor distributary; R_t = recharge due to seepage from tanks; R_{ci} = recharge due to percolation losses from areas irrigated by canals; R_{wi} = recharge due to percolation losses from areas irrigated by wells; R_{ti} = recharge due to percolation losses from areas irrigated by tanks.

All the components of eqns.(2) and (3) except Q_g were estimated using the recommendations of the Groundwater Estimation Committee of the Ministry of Irrigation (1984) with some modifications based on field observations, soils, crop and climate of the region. The following specific percentage allowance for various recharge components were used.

i) Recharge from rainfall (R_e) was taken as 17 per cent of the annual rainfall.

ii) Recharge due to seepage from major distributary (R_c), which is unlined and traverses a length of 53 kms was of the order of 3.0 m^3/sec./million m^2 of wetted perimeter (PPR, 1980). Hence, this value was used.

iii) A value of 1.75 m^3/sec/million m^3 of wetted perimeter (PPR, 1980) was used to estimate seepage losses from various minors (R_d). This was equivalent to 7 per cent of the discharge at the head of the minors.

iv) In the absence of data of water spread area of tanks, capacity (stored volume) criterion was adopted. Hence 20 percent of live storage (CPO, 1988) from tanks was taken as recharge due to percolation from tanks (R_t).

v) Return seepage from canal irrigated fields (R_{ci}) was 30 per cent.

vi) Tank irrigation system can be considered similar to canal irrigation system. Hence, 30 per cent value was assumed for estimating return seepage from tank irrigated areas.

vii) Recharge due to return seepage from well irrigated areas (R_{wi}) was assumed as 15 percent. This is because of better control exercised on this source of supply.

viii) Further, recharge due to deep percolation from wetland paddy crop fields, at a rate of 2mm/day per unit area for 120 days of the crop season was also considered.

The component, viz. subsurface inflow into/outflow across the aquifer boundaries (Q_g) was estimated as a residual from the detailed annual water balance study of the area. The following water balance equation was used :

$$(P+I) = (ET + E + Q_p + Q_g + Q_s \pm \Delta S_m \pm \Delta S_s \pm \Delta S_g) \qquad (4)$$

where: P = total precipitation over the area; I = irrigation water applied over the area from canal, tanks and wells ($Q_c + Q_t + Q_g$); ET = evapotranspiration; E = evaporation from uncultivated areas; Q_p = groundwater draft; Q_g = inflow/outflow across the boundaries; Q_s = Surface runoff; ΔS_m = Change in soil moisture storage; ΔS_s = change in surface water storage; and ΔS_g = change in groundwater storage.

All the above quantities were expressed in depth units mm (millimeter) over the gross command area of 83318 ha. Since the D-53 major was practically dry at the beginning of June every year (1983-84 to 1985-86), the change in surface water storage was assumed to be zero. Similarly, it was assumed that there was no significant change in soil moisture storage on annual basis. ΔS_g was calculated from the observation well inventory data (APSGWB, 1984-86) and specific yield of the aquifer. Average monthly rainfall data of the raingauge station located within the region are used to obtain P for the years 1983-84 to 1985-86. Monthly canal releases, quantity of tank water applied together with the total groundwater extracted was estimated by adopting suitable norms (Ministry of Irrigation, 1984) for different type of wells present were used to obtain I. ET was estimated from monthly data of potential evapotranspiration (Pan evaporation method) and the area under crops. Distinction was made between irrigated and unirrigated cropped areas. Evaporation from uncultivated areas (E) was calculated using Ritchie's equation (Ritchie, 1972) with suitable modifications. A value of 30 percent of normal rainfall was assumed to appear as surface runoff in the region (Sharma and Kampen, 1975, Krantz, 1979 and Virmani et.al. 1981) and it was suitably modified depending on actual annual rainfall during the period 1983-84 to 1985-86. The details how each component was estimated are given by Rao (1992).

Once all the components except Q_g in the eqn. (4) are known, Q_g was determined as a residual of eqn. (4) (Table I). The net annual recharge was estimated using eqns. (2) and (3). The resulting rise in groundwater levels was determined using the specific yield value of 0.125 for the years 1983-84 to 1985-86 (Table II). These values compared well with the observed values of water table rise which indicate that net regional groundwater recharge estimates were realistic.

Spatial distribution of net recharge : The estimated net recharge was distributed in space, amongst the different nodes over the flow domain based on the recharge distribution coefficients. At first the model was run with pumping only (no recharge allowed at the nodes). The difference between the model computed water levels and the observed groundwater levels at the end of one year period which were corrected for pumping effect was considered as the effect of recharge. The recharge distribution coefficients were then determined by distributing estimated net groundwater recharge weightedly in proportion to the difference of observed and model computed water levels at each node by the following equation (Sondhi et.al, 1989):

$$r_{ij} = H_{ij} - h_{ij} + Q_{ij}/AS \tag{5}$$

and

$$rd_{ij} = \frac{r_{ij}}{\frac{1}{MN} \sum_{i=1}^{M} \sum_{j=1}^{N} r_{ij}} \tag{6}$$

where: i, j = x, y coordinates of the concerned node; rd_{ij} = recharge distribution coefficient; r_{ij} = net groundwater recharge; H_{ij} = observed groundwater levels; h_{ij} = model computed levels with pumpage only; Q_{ij} = groundwater draft; A = area represented by node (i, j); and S = specific yield

The term $\sum_{i=1}^{M} \sum_{j=1}^{N} r_{ij}$ represents the total net groundwater recharge of the basin and MN represents the total number of nodes in the flow domain.

Table I : Annual Water Balance of the D-53 Command Area

Sr.No.	Water Balance Component	1983-84	1984-85 mm*	1985-86
Input				
1.	Rainfall (P)	1606.00	648.80	828.20
2.	Irrigation (I)			
	(a) Canal Supplies (Q_c)	405.96	473.16	302.80
	(b) Tanks	107.00	62.50	102.00
	(b) Wells (Q_p)	44.13	45.00	46.15
3.	Total Input	2163.00	1265.45	1279.15
Output				
4.	Evapotranspiration (ET)	223.10	250.40	238.80
5.	Evaporation (E)	571.00	442.00	454.00
6.	Pumpage (Q_p)	44.13	45.00	46.15
7.	Surface Runoff (Q_s)	803.00	171.00	207.00
8.	Increase in groundwater storage (ΔS_g)	103.15	89.25	59.40
9.	Total Output	1745.00	997.45	1005.35
10.	Groundwater Outflow (Q_g)	418.00	268.00	274.00

* Average depth over the gross command area of 83318 ha.

Table II: Annual Net Recharge to Groundwater in the D-53 Command Area.

Sl. No.	Source of Recharge	Volume of Recharge in MCM* 1983-84	1984-85	1985-86
1.	Recharge from rainfall (R_e)	227.50	97.00	117.30
2.	Recharge from D-53 major (R_c)	36.67	39.40	27.20
3.	Recharge from minors (R_d)	21.10	24.84	15.75
4.	Recharge from tanks (R_t)	17.83	10.41	17.00
5.	Recharge from canal irrigated areas including percolation from all irrigated paddy fields (R_{ci})	137.45	146.66	17.00
6.	Recharge from areas irrigated by tanks (R_{ti})	21.40	12.49	20.40
7.	Recharge from areas irrigated by wells (R_{wi})	5.51	5.63	5.76
8.	Gross Recharge (R_g)	467.50	336.30	317.00
9.	Groundwater Outflow (Q_g)	36.77	37.50	38.45
10.	Groundwater Draft (Q_p)	348.50	223.10	228.10
11.	Net Recharge (R_n)	82.20	75.70	50.36
12.	Equivalent rise in Water Table (m)	0.79	0.72	0.48
13.	Observed rise in Water Table (m)	0.82	0.71	0.47

MCM* = Million Cubic Metres.

Then, net recharge at each node (NR_{ij}) is given by

$$NR_{ij} = rd_{ij} * R_n \quad (7)$$

SENSITIVITY ANALYSIS AND VALIDATION OF THE SIMULATION MODEL

The reliability and quality of performance of simulation model was evaluated through sensitivity analysis. Separate simulations were carried out with the 1983-84 input data by varying the hydraulic conductivity (k) and storage coefficient (S) in turns. Hydraulic conductivity was varied by ± 25 per cent and storage coefficient by ± 50 per cent. The typical results of sensitivity analysis along the nodes diagonally across the D-53 command area (Rao, 1992) indicated that the model was particularly not sensitive to the values of `k' and `S'. Although the aquifer is considered to be homogeneous inspite of the spatial variations in the values of `T' and thickness of the aquifer, the initial assumed values of `k' and `S' seem to simulate closely the observed hydraulic heads. Hence initial values of `K' and `S' were retained for further analysis.

Since the time domain in the finite element simulation model was approximated using implicit finite difference scheme, it is unconditionally stable regardless of the size of the time increment. Convergence of the model was tested by incorporating the iteration closure (0.1m) criterion.

The model was validated against the historical observed data of groundwater levels for the year 1983-84 to 1985-86. The model computed groundwater levels and the corresponding observed values were used to draw the water table (Isobath) contour maps for the period 1983-84 to 1985-86. There is a close agreement between the observed and predicted water levels as seen from Fig.3. The average absolute difference between observed and predicted water-levels was 0.70 m, 0.311 m and 0.126 m for the years 1983-84 to 1985-86, respectively. Since the average absolute deviations between predicted and historical groundwater levels of 1m magnitude or more have been considered permissible (Klempt et.al., 1979 and Boonstra and DeRidder, 1981), it was, therefore, felt the model is capable of simulating the historic groundwater level reasonably well, and can be used for predicting the future water table behaviour in the region.

IMPACT ON GROUNDWATER REGIME

Based on the results of comparison study between irrigation water supply and requirement in the study area for the period 1983-84 to 1985-86, it was observed that the availability of water in the area was in excess of the irrigation water requirements for the cropping pattern envisaged in the original design. Further, this excess water led to rise in local groundwater levels (Bowonder and Ṙavi, 1983 and Rao, 1992). Thus the situation warrants proper conjunctive use of water resources in the command area of D-53 major distributary to arrest the rising groundwater table, as well as to increase the irrigation intensity thereby contrib-

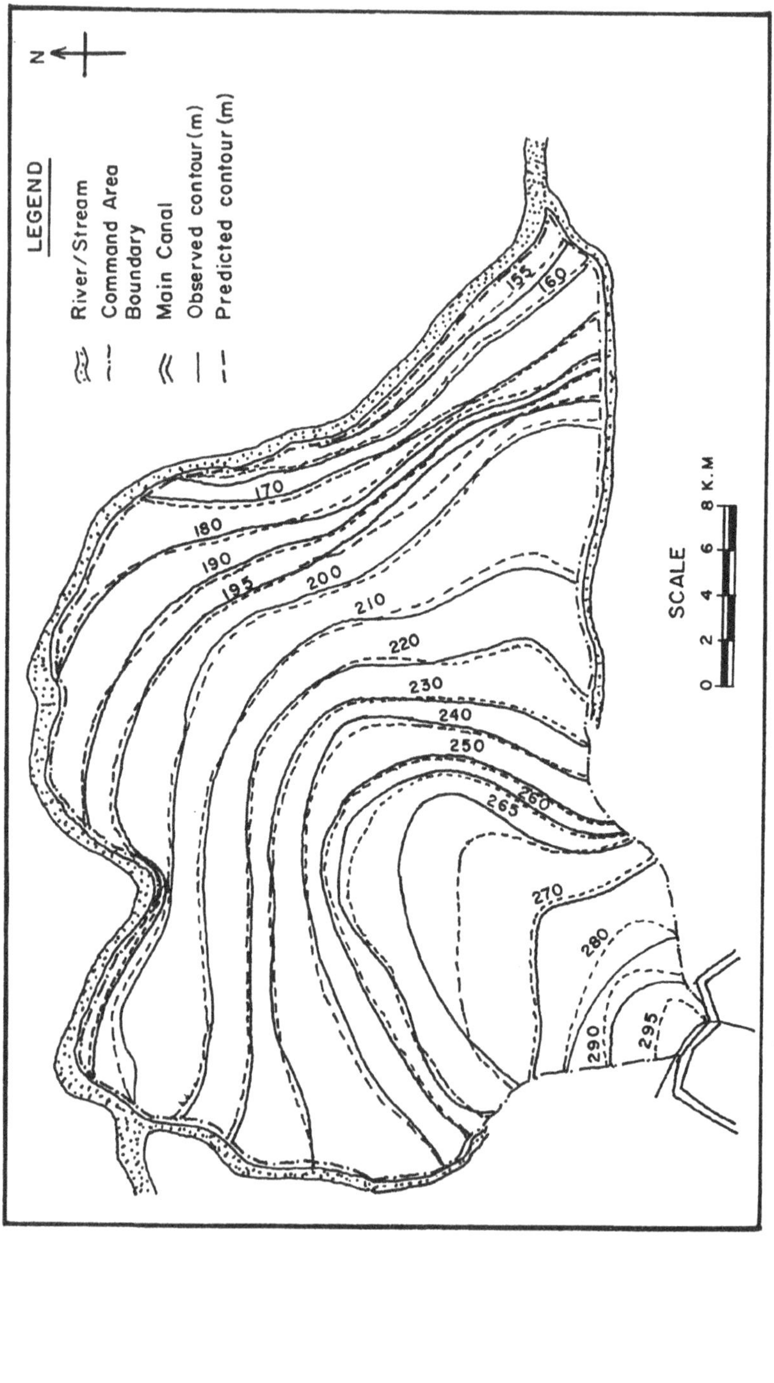

FIG. 3. OBSERVED AND PREDICTED GROUND WATER LEVELS (PRE-MONSOON, 1986) OF D-53 MAJOR DISTRIBUTARY COMMAND AREA.

uting to the increase in over all crop production in the area. Accordingly detailed assessment of water resources availability in the study area was made and the possible conjunctive water use plans were evolved (Rao, 1992). The validated groundwater simulation model was used to study the impact of these conjunctive use strategies on groundwater regime. In addition to this, the model was also used to determine the spatial distribution of additional groundwater potential and stabilizing pumping rates which may be made use in developing this additional groundwater potential.

CONCLUSION

The groundwater regime of D-53 major distributary of SRSP was simulated with a two dimensional finite element model using simplex triangular elements with linear trial functions. Net recharge to groundwater from different sources was quantified by conducting water budgeting, wherein norms of central groundwater estimates committee were suitably modified based on field data. The simulation model was successfully validated using the historic well inventory data for three years. Procedure for distributing the net recharge over the flow domain developed by Sondhi et.al., (1989) was successfully used in this study area. Since the model reproduces the past behaviour reasonably well, the recharge distribution coefficients determined from the procedure of spatial distribution of net groundwater recharge may therefore, be directly employed in future development of groundwater resources for conjunctive water use to arrest the rise of groundwater levels in the study area. The calibrated model was used to study the impact of a proposed crop-water management scheme on groundwater regime of the area wherein possible conjunctive water use plans were evolved.

During the course of this study, lack of adequate and reliable data necessitated various assumptions. Further, study was compounded by wide distribution of available data sources. Hence there is a need for groundwater data bases for effective formulation of the simulation model. The segment of the boundary which is not a natural boundary which has been considered based on well hydrograph records for three years of wells located in its vicinity as a constant head boundary need to be properly established either by field observation or well hydrograph records for longer period. Since this model is proposed to be used as an effective management tool for prediction purposes, the aquifer model needs to be updated and validated regularly.

ACKNOWLEDGEMENT

The authors are thankful to the Director, NRSA, Hyderabad and Head, Indian Institute of Remote Sensing, Dehradun for their continued support throughout the study and providing necessary help to prepare and present this paper. Authors are also thankful to Dr. N. H. Rao, Scientist, Water Technology Centre, I.A.R.I., New Delhi for his continued encouragement and guidance from time to time during the course of this study.

REFERENCES

APSGWB, 1984. Preliminary report on `Groundwater Development Possibilities'. District Series, Karimnagar, Andhara Pradesh.

APSGWB, 1984-86. `Records of observation wells in SriRamSagar Project', Andhra Pradesh, State Groundwater Board, Govt. of A.P., Hyderabad.

Baig, M.Y.A. and Prabhakar, S., 1980. `Analog Simulation of Unconfined aquifer in Pochampad ayacut', Andhra Pradesh. J. of Inst. of Engrs (India), **Vol. 61.** Pt. CI2.

Boonster, J. and DeRidder, N.A., 1981. `Numerical modelling of Groundwater basin. **ILRI Pub. No. 29**, Wageningen, Netherlands.

Bowonder, B. and C. Ravi, 1983. `Water logging in SriRamSagar Irrigation Project;. J. of Water Res. Dept. **Vol. 1, No. 3**, P 157-171.

CPO, 1984-86. `Mandal statistical reports', Office of the Chief Planning Officer, Karimnagar. Andhra Pradesh.

Klempt, W.B., Knowlees, T.R., Elder, G.R., and Sieh, T.W., 1979. `Groundwater resources and models applications for the Edwards (Balcones Fault zone) aquifers in the San Antonio Region'. **Report 239**, Texas Dept. of Water Res. Austin, Texas, USA.

Krantz, B.A., 1979. `Small watershed development for increased food production'. Res. bulletin, Int. Crops Res. Inst. for Semi-Arid Tropics, Hyderabad, Andhra Pradesh.

Ministry of Irrigation, 1984. `Report of the Groundwater Estimation Committee-groundwater estimation methodology', Govt. of India, New Delhi.

Murthy, K.N., 1991. `Sensitivity studies on water table aquifer parameters around Chelgal in SriRamSagar Command area'. J. of Indian Water Res. Soc., **Vol. II, No. 2**, 53-54.

Planning Commission, Government of India, 1985. Seventh Five Year plan, 1985-90, **Vol. II.**, Govt. of India, New Delhi.

Rao, V.V., 1992. `Studies on Water Management in Irrigation Canal Command areas - A case study'. Ph.D. Thesis, Div. Agric. Engg., Indian Agric. Res. Inst., New Delhi.

Ritchie, J.T., 1972. `A model for predicting evaporation from a row crop with incomplete cover'. Water Resour. Res., **8(5)**; 1204-1213.

Sondhi, S.K., Rao, N.H. and Sarma, P.B.S., 1984. `Assessment of Groundwater potential for conjunctive wateruse in large irrigation project in India'. J. of Hydrology, **1076**: 283-295.

Sreenivasulu, P., Jagmohan Das, G. and K.V.L.N.P. Choudhary, 1984. `Finite element analysis of inhomogeneous aquifers'. Proc. of the National workshop on the State of the art on `Water Resources Development and Management, Warangal, Andhra Pradesh: PP. 10.1-10.31.

Swain, L.A. 1978. `Predicted water-level and water-quality effects of artificial recharge in the Upper Coachella Valley, California, using a Finite Element Digital Model'. U.S. Geol. Survey WAter Reso. investigation, **77-129**.

Veeranna, M., 1990. `Report on Groundwater Resources and Development potential of karimnagar district', Andhra Pradesh. Central Groundwater Board, Ministry of water Resources. Govt. of India.

Virmani, S.M., M.V.K. Siva Kumar and S.J. Reddy 1982. `Rainfall probability estimates for selected locations of semi-arid India'. Res. Bulletin **No. 1**, ICRISAT, India.a

Water Technology Centre, 1983. `Resources analysis and plan for efficient water management - a case study of Mahi Right Bank Canal Command Area, Gujarat'. Indian Agric. Res. Inst., New Delhi, Res. Bull. **42**.

Water Science and Technology Library

1. A.S. Eikum and R.W. Seabloom (eds.): *Alternative Wastewater Treatment.* Low-Cost Small Systems, Research and Development. Proceedings of the Conference held in Oslo, Norway (7–10 September 1981). 1982
ISBN 90-277-1430-4
2. W. Brutsaert and G.H. Jirka (eds.): *Gas Transfer at Water Surfaces.* 1984
ISBN 90-277-1697-8
3. D.A. Kraijenhoff and J.R. Moll (eds.): *River Flow Modelling and Forecasting.* 1986 ISBN 90-277-2082-7
4. World Meteorological Organization (ed.): *Microprocessors in Operational Hydrology.* Proceedings of a Conference held in Geneva (4–5 September 1984). 1986 ISBN 90-277-2156-4
5. J. Němec: *Hydrological Forecasting.* Design and Operation of Hydrological Forecasting Systems. 1986 ISBN 90-277-2259-5
6. V.K. Gupta, I. Rodríguez-Iturbe and E.F. Wood (eds.): *Scale Problems in Hydrology.* Runoff Generation and Basin Response. 1986
ISBN 90-277-2258-7
7. D.C. Major and H.E. Schwarz: *Large-Scale Regional Water Resources Planning.* The North Atlantic Regional Study. 1990 ISBN 0-7923-0711-9
8. W.H. Hager: *Energy Dissipators and Hydraulic Jump.* 1992
ISBN 0-7923-1508-1
9. V.P. Singh and M. Fiorentino (eds.): *Entropy and Energy Dissipation in Water Resources.* 1992 ISBN 0-7923-1696-7
10. K.W. Hipel (ed.): *Stochastic and Statistical Methods in Hydrology and Environmental Engineering.* A Four Volume Work Resulting from the International Conference in Honour of Professor T. E. Unny (21–23 June 1993). 1994
10/1: Extreme values: floods and droughts ISBN 0-7923-2756-X
10/2: Stochastic and statistical modelling with groundwater and surface water applications ISBN 0-7923-2757-8
10/3: Time series analysis in hydrology and environmental engineering
ISBN 0-7923-2758-6
10/4: Effective environmental management for sustainable development
ISBN 0-7923-2759-4
Set 10/1–10/4: ISBN 0-7923-2760-8
11. S.N. Rodionov: *Global and Regional Climate Interaction: The Caspian Sea Experience.* 1994 ISBN 0-7923-2784-5
12. A. Peters, G. Wittum, B. Herrling, U. Meissner, C.A. Brebbia, W.G. Gray and G.F. Pinder (eds.): *Computational Methods in Water Resources X.* 1994
Set 12/1–12/2: ISBN 0-7923-2937-6
13. C.B. Vreugdenhil: *Numerical Methods for Shallow-Water Flow.* 1994
ISBN 0-7923-3164-8
14. E. Cabrera and A.F. Vela (eds.): *Improving Efficiency and Reliability in Water Distribution Systems.* 1995 ISBN 0-7923-3536-8

Water Science and Technology Library

15. V.P. Singh (ed.): *Environmental Hydrology*. 1995 ISBN 0-7923-3549-X
16. V.P. Singh and B. Kumar (eds.): *Proceedings of the International Conference on Hydrology and Water Resources* (New Delhi, 1993). 1996
    16/1: Surface-water hydrology ISBN 0-7923-3650-X
    16/2: Subsurface-water hydrology ISBN 0-7923-3651-8
    16/3: Water-quality hydrology ISBN 0-7923-3652-6
    16/4: Water resources planning and management ISBN 0-7923-3653-4
    Set 16/1–16/4 ISBN 0-7923-3654-2

Kluwer Academic Publishers – Dordrecht / Boston / London

SPRINGER NATURE

GPSR Compliance

The European Union's (EU) General Product Safety Regulation (GPSR) is a set of rules that requires consumer products to be safe and our obligations to ensure this.

If you have any concerns about our products, you can contact us on ProductSafety@springernature.com

In case Publisher is established outside the EU, the EU authorized representative is:

Springer Nature Customer Service Center GmbH
Europaplatz 3
69115 Heidelberg, Germany

The manufacturer's authorised representative in the EU is Springer
Nature Customer Service Centre GmbH, Europaplatz 3, 69115 Heidelberg,
Germany. If you have any concerns regarding our products, please
contact ProductSafety@springernature.com

Printed and bound by CPI Group (UK) Ltd, Croydon, CR0 4YY
30/09/2026
02225436-0005